HOW TO BE MULTIPLE

BY THE SAME AUTHOR

Artful Truths: The Philosophy of Memoir

HOW TO BE MULTIPLE

THE PHILOSOPHY OF TWINS

HELENA DE BRES

ILLUSTRATIONS BY JULIA DE BRES

BLOOMSBURY PUBLISHING
NEW YORK · LONDON · OXFORD · NEW DELHI · SYDNEY

BLOOMSBURY PUBLISHING
Bloomsbury Publishing Inc.
1385 Broadway, New York, NY 10018, USA

BLOOMSBURY, BLOOMSBURY PUBLISHING, and the Diana logo
are trademarks of Bloomsbury Publishing Plc

First published in the United States 2023

Copyright © Helena de Bres, 2023

Illustrations copyright © Julia de Bres, 2023

All rights reserved. No part of this publication may be reproduced or transmitted in any form or by any means, electronic or mechanical, including photocopying, recording, or any information storage or retrieval system, without prior permission in writing from the publishers.

Illustration on page 2: John Tenniel's illustration for chapter 4 of Lewis Carroll's *Through the Looking-Glass*, originally published 1871. Public domain.

Bloomsbury Publishing Plc does not have any control over, or responsibility for, any third-party websites referred to or in this book. All internet addresses given in this book were correct at the time of going to press. The author and publisher regret any inconvenience caused if addresses have changed or sites have ceased to exist, but can accept no responsibility for any such changes.

ISBN: HB: 978-1-63973-034-6; EBOOK: 978-1-63973-035-3

LIBRARY OF CONGRESS CATALOGING-IN-PUBLICATION DATA IS AVAILABLE

2 4 6 8 10 9 7 5 3 1

Typeset by Westchester Publishing Services
Printed and bound in the U.S.A.

To find out more about our authors and books visit www.bloomsbury.com and sign up for our newsletters.

Bloomsbury books may be purchased for business or promotional use. For information on bulk purchases please contact Macmillan Corporate and Premium Sales Department at specialmarkets@macmillan.com.

For Julia—my second, and my first

CONTENTS

Twins in Wonderland — 1

1. Which One Are You? — 27

2. How Many of You Are There? — 71

3. Are You Two in Love? — 113

4. How Free Are You? — 153

5. What Are You For? — 187

Coda — 229

Author's Note — 233

Acknowledgments — 235

Notes — 239

HOW TO BE MULTIPLE

TWINS IN WONDERLAND

My favorite picture of twins is John Tenniel's illustration of Tweedledum and Tweedledee for Lewis Carroll's *Through the Looking-Glass*. The Tweedles stand beside each other in a dark wood wearing matching school uniforms, one with DUM embroidered on the collar, the other with DEE. Alice leans toward the pair, captivated, while they warily return her gaze, each with one arm around the other's back.

The scene deftly captures three core aspects of identical twinhood. For one thing, it features two people who look strikingly alike. Since Leibniz published his *Discourse on Metaphysics* in 1686, philosophers have held that no two distinct things can *exactly* resemble each other. ("If it was so," says Tweedledee on another subject, "it might be; and if it were so, it would be; but as it isn't, it ain't. That's logic.") The Tweedle brothers must differ, then, but how? Who could resist peering closer at these slippery characters, in search of clues to what makes one Dum, the other Dee, beyond the names stitched on their collars?

The picture also underscores the closeness that most of us assume twins share. Tweedledum and Tweedledee are positioned so near each other they look physically conjoined. Emotionally, too, they're clearly intimate and aligned: check out the matching side-eyes they shoot in Alice's direction and their protective-possessive side-hug. But the image foreshadows that things between them aren't as peaceful as they seem. Lying at the twins' feet is an umbrella they'll later use as a weapon, when a petty disagreement between them escalates into a violent battle. Twinship is always passionate, that umbrella reminds us, but where the passion will lead is anyone's guess.

Finally, though the twins are the alleged stars here, it's Alice's reaction to them that the illustration foregrounds. While she's mobile and lively on the left, Tweedledum and Tweedledee are static and stolid on the right, like a pair of inanimate objects awaiting her eager examination. Much of the experience of twinhood is determined not by twinship itself but the response of non-twins to it. Enthrallment can be gratifying and twins often encourage it, as Tweedledum and Tweedledee later do, when performing a poem for their audience of one. But, like all twins, these two know that attention from a singleton always merits their suspicion.

Part of what I love about this picture is that, by including each of these elements, it captures the experience of twinhood from the inside and the out. I am an identical twin, so I mainly identify with those tubby schoolboys, hugging each other affectionately and watching Alice watching them. But I also identify, like the standard viewer does, with Alice. We twins are as susceptible as anyone else to the multiple charms of twins.

• •

Tweedledum and Tweedledee live in Wonderland, a place peopled by the odd, unpredictable, seductive, and unhinged. Uncoincidentally, wonder is also where Socrates claimed philosophy begins. It's no great surprise, then, to find some deep existential questions lurking in the vicinity of the Tweedle twins and their counterparts here aboveground.

What makes a person themself rather than someone else? Can one person be spread across two bodies? Might two people be contained in one? What distinguishes perfect from pathological love? Can our actions be truly free, when so much about us is beyond our control? At what point does wonder at another turn into objectification? ("If you think we're wax-works," Tweedledum tells

Alice, "you ought to pay, you know.") We twins vividly raise these questions by the bare act of existing, and much of our power to fascinate springs from them.

As well as being a twin, I'm a philosophy professor, and for most of my life I haven't considered those parts of myself in tandem. My twinhood felt like something scientists or psychologists might pronounce on or artists and cultural theorists illuminate, not something related to my own research and teaching. But the instant the possibility occurred to me, I was flooded by the connections between my knowledge of philosophy and my experience of twinhood, like Alice surprise-drowning in the sea of her own tears. And I also sensed a great, yet-to-be-explored conceptual and personal territory, like the luxuriant garden behind the tiny door that Alice opens with her magic key.

Tweedledum and Tweedledee are bit players in Alice's adventure, curiosities not to be tarried with for long. What would it be like to center them on the journey through Wonderland, I asked myself, and follow through on the mysterious questions they raise? And what would it be like to do that from the inside of the experience—keeping my life with my own twin, Julia, always in view? So many of our cultural ideas about twins are driven by the perspectives and priorities of singletons. Given the numbers, that's maybe inevitable, but it's an unfortunate restriction on humanity's intellectual resources. Don't you Alices out there wonder occasionally what we Tweedles are thinking and feeling—about ourselves and about you—as we trip back into our forest?

• •

This book, a collection of linked personal essays on the obscure landscape behind Tweedledum and Tweedledee, is the result of that intimate adventure in Wonderland. It uses my experience of twinhood and my background as a philosopher, combined with my further discoveries in scholarship, art, and popular culture, to explore the assumptions underlying our intense reactions to twins and show how they illuminate wider questions about what it means to be human.

The first essay narrows in on the thorny question of what makes Tweedledum Tweedledum and Tweedledee Tweedledee. When twins aren't treated as indistinguishable, they're often cast as binary opposites, on one or more axes. One twin's the Chaos Muppet, for instance, the other the Order Muppet; one's the empath, the other the narcissist; one's the tomboy, the other the femme. This tendency to binarize twins is rife in individual families, and in myth, art, and culture across the board. Twins are such a lightning rod for binary thinking that they're often used as avatars for other salient binaries in debates over politics and the good life, and then those cultural avatars are used to reinforce the binarization of real-life twins.

What explains this widespread habit of dumping twins into two sharply opposed camps? Is there anything good to be said for it, alongside the obviously bad things? And once you realize that your sense of yourself has been formed by binary thinking—whether you're a twin or not—how should you feel about that? This essay uses the special instance of twins to think about the general motivations and costs of binarization as a human tendency, and what it says about the nature of the self.

The second essay is about another question of personal identity that twins forcefully raise. A flyer for an 1869 show in London by Chang and Eng Bunker, the original "Siamese Twins," read:

> ARE THEY TWO MEN? Numerically they are two; but with men of science it is a disputed point, if there is more than one *person!* Is it one *man* in two bodies, with a double set of limbs and organs? Is there but one life, one responsible will, one accountable being, only with double bodies, double heads, and a double set of arms and legs? Or, are they really two men—each distinct from the other as Smith from Brown, with the exception of that one mysterious bond?

This question of "the mathematics of personhood"—how to draw the boundaries between one person and another—comes up most obviously in the case of conjoined twins, but extends to the non-conjoined kind, too. In everyday life, all twins are susceptible to being treated as, if not the same person, at least somewhat less than two. Cultural representations of twins often imply that twins share a self, somehow split between or duplicated across them.

I used to think the idea that twins share a self was obviously false, but I'm no longer so sure. Working out whether it's false requires asking larger questions that apply to singletons as much as twins. Where does the core of you reside—in your body or your mind? And is there any sense in which either of those things could extend beyond the skin or skull we ordinarily associate with you, so your personhood overlaps with another human? Twins vividly breach some of the central physical, cognitive, and emotional boundaries we assume hold between individual people. Thinking about their case can help us think about the more general human case, with far-reaching implications.

One reason twins seem unusually interconnected is the intense emotional bond they often share. The third essay turns directly to the relationship between twins and love, which provokes complex reactions. For many people, twins included, twinship represents the ideal human attachment, marked by a degree of closeness and affection of which singletons can only dream. In the extreme version of this fantasy, twins enjoy complete equality, deep intimacy, constant companionship, selfless mutual concern, magnified powers, and the happiness that comes from living with your soulmate. But alongside this positive vision of twinship we see a much darker one. Literature and popular culture are rife with morbidly entwined, often incestuous, twins who become ever more isolated, then sink into stagnation and decay. Alternatively, twins are portrayed as perennially at each other's throat—sometimes literally. A weirdly large proportion of fictional narratives about twins end in murder, suicide, or mental breakdown.

Singleton portrayals of twins often imply that, while twin love is enviable in youngsters, twinship is something you should *grow out of.* The ideal adult relationship, the suggestion goes, is between people who are different, and who join those differences, without eliminating them, into a new, generative whole. This essay unearths the assumptions built into that view—which go to some dark places—with the goal of exploring how each of us might benefit from widening the set of loving relationships we see as healthy, fulfilling, and meaningful.

The fourth essay turns to the relationship between twins and free will. Twins trigger many singletons into doubting the existence of human free agency, sometimes to the point of existential panic. The likeness of identical twins goes well beyond their looks, to encompass their interests, values, talents, and personality traits. If we assume that shared genes explain these similarities, it looks as if much of who any of us are derives from biological forces beyond our control. If who we are determines in turn how we behave, how can any of our actions be genuinely chosen? Identical twins whose lives take very different paths inspire a somewhat different anxiety. Like the two characters Gwyneth Paltrow plays in the rom-com *Sliding Doors*—which splits into parallel story lines partway through, narrating the dramatic consequences of Paltrow's either catching or not catching a single train—twins with strikingly diverse experiences underscore for singletons the role that purely circumstantial factors play in how their own lives turn out.

My anecdotal impression is that we twins are much less worked up about all this than non-twins are. This essay takes on the philosophical debate over free will to consider why that might be so, and what the perspective of twins might tell us about the kinds of freedom that really matter.

The final essay explores the relationship between twins and objectification. When they're together, identical twins with similar appearances are permanent magnets for what we might call "the singleton gaze." Even usually sensitive and well-meaning people stare at us, obsess over our appearances, and ask us and our parents invasive questions. There's the weirdly popular "Who's older?"—answer: "Who cares?!"—but also "Are you breastfeeding them?"; "Which one's naughtier?"; "Who's the smarter one?"; and "Why are you skinnier?" Singletons also use us for their own ends in a wide variety of ways, in science, art, entertainment, and the market. Much of this seems objectifying, but it's not clear what that really means, or how worried we should be about it.

An appealing thought here is that what matters is how complicit the objectified are in their own objectification. If, like Tweedledum, we ask our Alices to pay and they do, what's the problem? Maybe the objectified can reclaim their agency and dignity by playfully subverting the very tropes that belittle them. If so, even extreme cases of twins stoking up clichés about themselves for singleton amusement shouldn't trouble us.

I have mixed feelings about this line of thought. This essay parses them out, using the case of twins to get clearer on more general questions about the ethics and politics of othering your fellow humans.

• •

While each of these essays can be read separately, they share a theme, one that only came to light as I started to write and talk over what I was writing with Julia. Much of the philosophical interest of twins, I've come to think, derives from the challenge they pose to our usual understanding of how "normal" humans look, think, feel, act, and relate.

In industrialized countries the twin rate has roughly doubled since 1980, due to the rise of assisted reproductive technology and the average age of pregnancy. But twins remain rare: in the United States and Europe today, only one in sixty people is a twin. Of these, even fewer are identical twins: we appear in only four out of a thousand births (a rate that applies to every mammal, apparently, except the overachieving armadillo).

We twins aren't *simply* rare, as, say, people with green eyes are. Instead, in subtle and not-so-subtle ways, we're perceived as alien, as "other," by the singleton majority. To be clear, this othering reaches nothing like the oppression faced by racial and gender minorities, disabled people, and women. There's no widespread asymmetrical power structure with singletons at the top and twins at the bottom. Twins aren't at greater risk of violence or abuse than non-twins (at least as adults; they're at somewhat greater risk when young). They don't suffer unequal opportunities in education, employment, and political participation; they don't face the kind of deep-seated, unrelenting social stigma that makes everyday life a struggle.

Still, from the majority perspective, we twins are *weird*, and that has consequences. For a start, we're hypervisible: we attract double takes and invite stares. In nineteenth- and twentieth-century freak shows, we were displayed as part of the "congress of human oddities" (as one sideshow banner put it) alongside others with unusual anatomies. Whatever mystique our doubled forms provide tends to be sullied by the assumption that we're somehow defective. Conjoined twins are treated as manifestly physically inferior or dysfunctional, but the literature in obstetrics and psychology takes a paranoid position about even separate twins, assuming we're at exaggerated risk for a variety of ailments and neuroses. Alessandra Piontelli, a specialist on twin development, calls this the twin "neo-gothic": a voyeuristic obsession with traumatic births, codependence, and dysfunctional extratwin relationships.

It's not only our bodies, but also our hearts and souls that are under suspicion. As Dostoyevsky's character Petrushka put it, "Good people live honestly, good people live without any faking, and they never come double . . . Yes sir, they never come in twos, and they're not an offense to God and honest people." Singletons across history have seen us as shifty, untrustworthy, at perpetual risk of signing up for the dark side. In many cases, alongside our virtue our very humanity has been cast into question. Twins are sometimes represented as subhuman: compared to animals born in litters. More often, we're seen as *super*human: connected somehow to the supernatural, though not often in a good way. Sometimes we're regarded as powerful, without being malevolent. We've been treated as divine beings, lucky charms, and good omens. We've been claimed to have a direct dial to God, to be able to predict the future, and to detect the workings of thieves. The Tsimshian people of British Columbia believe we're in control of the elements: "Calm down, breath of the twins," they pray, when the skies gust or pour. A lot of the time, though, our association with the supernatural inspires horror or fear in singletons. Twins have been interpreted as the sinful products of adultery or demonic possession and linked with the uncanny, sinister, and evil. In our Gothic doppelgänger form, we're seen as ghostly premonitions of moral downfall or death. In some cultures, men have cut off one of their testicles in an effort to avoid fathering us, and women receive the terrible curse "May you become the mother of twins!"

As a result of our supposed deviance, twins are subtly policed. We're meant to be alike, but not too alike. We're meant to love each other, but—whoa, there!—not too much. We're meant to retain a constant, invariant bond, but also to grow up. Though these everyday forms of social pressure are mainly just annoying, other singleton treatments of twins reach the level of material disrespect and maltreatment. Purely as a result of their twinhood, twins have been exiled from their villages and sold into slavery. They've been coerced to sleep with each other and subjected to involuntary scientific experiments at great physical and emotional cost. And they've been murdered at birth, violently assaulted, or raised solely to be sacrificed to the gods.

Such appalling violations are the exception. Many particular twins enjoy major social privilege, and twins in general are often treated as special in beneficial and welcome ways, just for being twins. Our case strikes me as an arresting example of the truly impressive reach of the human tendency to other. That we social darlings are subject to an array of everyday prejudices, simply due to our innocuously coming in twos, underscores the narrowness of the line that all of us—twins or not—are pressured to walk.

∙ ∙

Twins are *freaks*, in the nonpejorative dictionary sense: we're odd, unusual, anomalous. Many types of people fall into that category, but in writing this book, I've begun to think the freakishness of twins is special. We twins don't simply depart, in our twoness, from the singleton norm. Instead we're *superfreaks*: symbolic, in certain ways, of all who step outside society's rigid strictures about the right way to be and act.

Elizabeth Grosz notes that most monsters throughout history have been creatures of excess, "with two or more heads, bodies or limbs, or with duplicated sexual organs." This suggests that to be a freak is often to be *too much*: to spill voluptuously and licentiously outside the bounds of normality and decency. On the freak show circuit, for instance, the Hottentot Venus had too large an ass and breasts, the Bearded Lady too much hair, the Four-Legged Lady too many limbs, and the Fat Lady too much flesh.

Alice herself becomes one of the freaks of Wonderland when her supersized body threatens to explode through the rafters of the White Rabbit's house.

A related characteristic of many monsters and freaks is a disturbing hybridity, a union of things we ordinarily take to be separate. Griffins, centaurs, sphinxes, and basilisks ignore the distinction between lion and raptor, horse and human, serpent and fowl. This is another way of being too much—things are meant to be one thing, not several—and produces a similar sense of spillage and disorder. A popular freak show figure was the half man / half woman, dressed like a gentleman on one side and a lady on the other. Conjoined or parasitic twins (or, as they used to be called, double monstrosities) were among the sideshow's most treasured attractions.

What's more too much than a multiple birth? Each twin is problematically duplicated: there are two where there ought to be one. And the emotional and cognitive boundaries between twins, whether the pair are physically conjoined or not, seem to singletons disturbingly permeable, hybrid-like. Twins are a walking instance of the essence of freakdom, a metaphor for the rest of the team.

Grosz attributes the awe and horror freaks inspire to the way their excess and hybridity violate the basic categories we use to understand our experience. Freaks disturb our familiar binary oppositions between male and female, Black and white, primitive and cultured, child and adult, animal and human, normal and abnormal, self and other. They destabilize our grip on the world and others and ignite a fear that we might lose hold of our own identities, to the point that we get submerged in, fused with, or incorporated into an alien other. This prospect horrifies many of us raised in Western cultures because we're deeply attached to being discrete individuals with sharply distinct physical, mental, and emotional barriers between ourselves and other humans. Our ideal of a healthy and dignified person is an adult in full control of their mind and body, who rationally pursues their personal aims largely independent of the influence and interference of others. But the possibility and desirability of achieving that ideal is suspect. Twins and other freaks make visible our physical and emotional vulnerability and our social interdependence and model more communal ways of living. They fail to uphold the natural order of human division.

The social role of the traditional freak show, as Rosemarie Garland-Thomson has argued, was to reassure audiences of their own normality, beauty, and virtue. Self-doubting able-bodied slim white straight singletons of average height, ass size, and hairiness could come away from the spectacle with a renewed sense of the border between the ordinary and the aberrant, and their secure place on the correct side of it, a pretty good deal for a dollar. In a subtler way, twins continue to offer this service for free. They act as a set of dark twins to the "normal" human, defining and validating that person by way of opposition. Opposed to the singleton viewer *as* twins, they can confirm that singleton as an appropriately discrete individual. But they can also serve as proxies in the theater of the singleton's imagination for any number of other oppositions between the singleton and other social groups. One twin is often cast as the strong, the other the weak; one the rational, the other the emotional; one the civilized, the other the savage. In each case, singletons can align themselves symbolically with the twin they prefer and cast the other twin, and all it represents, into darkness. Jung notes that in many languages *second* and *other* are simply the same word.

• •

Generalizing about twins is risky. For one thing, as many people instantly told me when I mentioned I was writing this book, you don't want to lose sight of the important differences between *identical* and *fraternal* twins. Those terms are used in everyday language to distinguish between twins who result from a single egg that splits after fertilization and twins who develop from two separate eggs that are fertilized around the same time. Those born from a single egg—referred to by scientists as monozygotic or MZ twins—are almost genetically identical. (They start with the same DNA, but minor changes in DNA may happen over time, and in any case the same genes can be expressed differently.) Those born from different eggs, dizygotic or DZ twins, share no more genetic material than your standard pair of siblings.

I like the terms *single-egg* and *different-egg* twins better than their *identical* and *fraternal* alternatives, partly because they're more informative, partly because I've had a lifetime of people asking if Julia and I are identical, one of us confirming that we are, and then our questioner objecting, as if personally wronged, "But you look different!" *No shit, Sherlock.* Still, *identical twins* is what we in the single-egg camp usually get called, so much of the time, including here, I just play along.

Most people see identical twins as the paradigm of twinhood, despite their being the much-rarer type. The large majority of representations of twins in contemporary culture feature identicals, and while few would go so far as the French, who refer to fraternal twins as *faux jumeaux*—ouch!—who among us doesn't feel a secret twinge of disappointment when someone reveals they're a twin and then disambiguates in the nonidentical direction?

I'm suspicious of the headlining of identical twins because I suspect it springs from an obsessive focus on twins' physical appearances, at the cost of the relational dimensions of twinhood that tend to matter most to twins themselves. All twins who are raised together, identical or not, share growing up in step with another sibling from conception. All such twins, identical or not, have to deal with the reactions that unusual phenomenon elicits from the rest of the world. If the heart of twinship is that kind of social experience, I see no reason to demote my fraternal brethren to a lower level of twinhood. Still, because much of this book is concerned with exploring our dominant cultural image of twins, as well as my own experience of identical twinhood, the imbalance is hard to avoid. I'll often refer to *twins* generally, for the sake of style, but I think it'll usually be clear when I have only identical or only fraternal twins in mind.

Another overgeneralization I risk here is taking my own experience of twinhood as emblematic of twin experiences, period. We twins are a superdiverse lot, and one of our pet peeves about singletons is their habit of homogenizing not only the members of our particular twinship, but also twins as a group. I've tried not to do that second thing, but can't promise I've succeeded, and if some unjustified twin-essentialism slips in, I apologize. Fundamentally, the essays here are about one experience and understanding of twinhood, mine. That said, like many a personal essayist, I've written them with the sneaky conviction that the picture served up will have broader reach.

This question of what is and isn't special about twins leads to a more general one, relevant to everything this book represents. When it comes to philosophical questions of personal identity, love, freedom, and the like, are twins a unique case, an enlightening exception to the human norm? Or are twins intellectually illuminating, instead, because they offer particularly vivid examples of experiences all humans share? (This second option reminds me of the section in *US Weekly* that shows celebrities looking disheveled or hungover, with the heading STARS: THEY'RE JUST LIKE US!)

My philosopher friend asked me this question recently, and I didn't quite know how to answer. On the one hand, part of what interests me—and everyone—about twins philosophically is the light they promise to shed on human experience in general. Many of the aspects of twinhood I'll talk about have clear parallels with non-twin experiences. Twin relationships share much with all sibling relationships, for instance, and with other intimate bonds. I'll also be exploring intersections between my and Julia's experiences of being twins and being disabled, queer, and female. On the other hand, some aspects of twinhood read to me as genuinely unique, and in my more mystical moments I'm tempted by the idea that they generate a kind of secret knowledge that may have little bearing beyond their own case.

I think both things are true: twins are in some ways special, in others not. I'll emphasize one or the other theme at different points, depending on the topic at hand and my perspective on it. The persistent tension between sameness and difference, the desire to have your similarity recognized as well as your uniqueness preserved, isn't exactly new to me in life. If, in oscillating between these takes, these essays embody that twin-like ambivalence, then maybe so much the meta.

• •

I was standing at a circulation desk with Julia last year, checking out a book called *Freaks*, when the librarian noticed the title.

"What are you reading *this* for?" she asked me.

"I'm writing a book about twins." I replied.

"Twins aren't freaks!" the librarian exclaimed. "I'm a twin!"

"So are we," I noted, gesturing at Julia.

The librarian set off on a string of reminiscences about her and her sister, while Julia and I listened attentively. Occasionally I'd shoot a side glance at Julia.

Freak, I'd silently comment, in reference to our new friend.

Yep, she'd beam back. *Freak*.

Not all twins recognize themselves as freaks, and still less do the more privileged singletons among us. Many react to the idea the way Alice does, when the Cheshire Cat suggests she can visit either the Mad Hatter or the mad March Hare. "But I don't want to go among mad people," Alice remarks primly, librarian-style. "Oh, you can't help that," replies the Cat. "We're all mad here. I'm mad. You're mad."

The fear that this might be true—that each of us is, if not mad, at least odd, deviant, in violation of the norms we've absorbed since infancy about the proper ways to act and to be—is central to the fascination twins, among other freaks, exert. Carson McCullers's character Frankie Addams expresses it when she visits a Freak Pavilion in *The Member of the Wedding*. She feels the people on display "looked at her in a secret way and tried to connect their eyes with hers, as though to say: *we know you!*" "Do you think I will grow into a Freak?" she asks uneasily.

Too late, Frankie. The social boundaries freaks violate are ones that everyone violates or risks violating in some way. Only a minority of people satisfy all of our culture's demands, and their success there is always unstable. We're each only a car accident, a medical diagnosis, a queer crush, a religious conversion, or a genealogy certificate away from mainstream social failure. In light of this, it'd be better to just accept, as James Baldwin writes, that "we are all androgynous . . . each of us, helplessly and forever, contains the other—male in female, female in male, white in black and black in white." But "the human being does not, in general, enjoy being intimidated by what he/she finds in the mirror."

Twins function as the mirror, the dark side of the Looking-Glass. In reflecting each other, they reflect the rest of us. As McCullers's Frankie says later in the book, of her sister and future brother-in-law, they are "the we of me."

Acknowledging yourself in the other is one way of signing up for a freer and more just society, but not the only way. We could also come to recognize others as very different from us, and not for that reason any less valuable or deserving of our respect. But whether we consider freaks similar to or different from us, thinking about those who fall outside the norm has liberating potential for everyone. Being normal takes a lot of work. Twins and other freaks offer images of forms of living that are more capacious and less oppressive than those our society has taught us to value: new, freeing ways to think, act, and feel.

That's also one of the traditional roles of philosophy, so twins and philosophy are a natural fit. Well—that might be truer of some particular twins than others.

Julia and I took the same introductory philosophy course in our first semester of college; she got an A and I got an A+. In an instance of the magnification of minor differences that often happens with twins, Julia decided she was terrible at philosophy and never took a class in it again. I told our professor this at a party a few years later.

"Oh my God," he gasped, abruptly setting down his beer. "I was probably just undercaffeinated when I was grading her final. You were both so good at it! It meant nothing!"

I could see he felt like some kind of hapless god in that moment, capable of flicking the switch on the train tracks so that two hitherto identical life courses permanently and consequentially diverged. Maybe the horror I saw in his eyes also reflected that more personal metaphysical panic twins often inspire in singletons. Was his own life determined by meaningless accidents in this same way? Whose lack of coffee had fucked *him* up?

He shouldn't have worried, at least not about the first part. Julia's quitting philosophy was overdetermined: good at it or not, she hated it in college; it made her eyes glaze over. "Omg, what is the point? What is even the *point*?" she'd wail as we did our course readings together on the couch. She's more open to it now, but still approaches philosophical theory as if it's on borrowed time and has to prove its value to her pronto. You'd think I'd find that annoying, but, as with many of our differences, I love it, and it's useful, too. I got Julia to read this book many times and filter out the boring parts. (The ones that are left are her fault now.) We also decided she'd draw pictures for it, and her expertise in analyzing social discourses, the research area she went for after ditching philosophy, has informed a lot of what's inside. We're hoping the result will open your eyes and be a good time, too. After all, isn't that what we expect from an encounter with twins in Wonderland?

1
WHICH ONE ARE YOU?

When my identical twin came to visit me in Boston recently, I took her to my favorite café for a coffee and a waffle. As we settled into our drinks, I glanced at the barista, a lavishly tattooed heteroflexible twenty-seven-year-old with an unsympathetic landlord, a crazed ex-girlfriend, and a complicated relationship to Orthodox Judaism.

"That guy has a pet rabbit," I whispered.

Julia looked up instantly and beamed.

"I hear you have a rabbit!" she shouted.

I've spent many hours writing in this café over the years and have picked up these biographical tidbits in the ether, but I haven't spoken to this barista for longer than it takes to say, "A latte and a waffle, thanks." Once a season or so, if I'm pre-caffeinated and feeling wild, I'll add, "Crazy weather we're having, huh?" then swiftly retreat, made slightly breathless by my own extravagance.

The barista was enveloped in a cloud of steam from the espresso machine. He peered out of it, uncertainly.

"Ye-es," he said.

"What's its name?" Julia asked brightly.

"Bertha," the barista answered. "Wait a second."

Then the barista was at our table, walking Julia through Bertha's Instagram page and discussing the skidding issues large bunnies face on hardwood floors.

"I left home at seventeen and have lived alone ever since," said Bertha's dad. "Bertha is the first time I've ever had to look after anyone other than myself. It's taught me to be more responsible, you know? Even in a financial sense."

Julia nodded. "I totally get that," she said, leaning in. "Aww! Look at her fluffy paws! Bertha, you little rascal, you!"

When people ask how my twin and I differ, this is the kind of example I offer. For as long as I can remember, I've known that Julia is the extrovert and I'm the introvert, though I wouldn't have put it in those terms when we were kids. Back then I'd have described myself as "quiet" or "happy by myself" or "hard to get to know," and Julia as the opposite. It's part of the family mythology that this contrast was apparent from day one, as we lay in our incubators at St Helens Hospital in Auckland, New Zealand, across a golf course from what's now called the Twin Coast Discovery Highway. We'd been expected around Christmas, but, like many twins, arrived a month early, weighing just a few pounds each. When our parents visited us in the ICU, tense with excitement and anxiety about their tiny, fragile creations, they found me wrapped contentedly in my swaddling cloth, calmly observing my environs like a philosophical burrito. Julia had worked her way out of her confines and was energetically smearing her shit onto the Perspex walls, making eyes at the nurses and crowing with delight. "Hi, everyone!" our mother reports her nonverbally beaming. "Let's get this party started!"

A lot of traffic has moved down the Twin Coast Discovery Highway since Julia and I were born, but this basic contrast hasn't hugely shifted. While I flee from committees, Julia establishes them, for fun, and asks everyone on them out for coffee. Though I'm no recluse, I conserve my social energies for a select few, and I like to keep my inner life on, you know, the *inside*, where it belongs.

Many twins have some version of this pair portrait ready to hand out on demand. The specifics differ— maybe the central contrast isn't between the introvert and the extrovert but between the leader and the follower, the conformist and the free spirit, the athlete and the aesthete, the clown and the sage. Still, the basic structure insistently recurs: two clearly differentiated options, with one twin squarely inhabiting each.

We find this general setup in mythological traditions all over the planet. Egypt has Nut, the star-spangled goddess of the sky, and her twin brother, Geb, the viper-headed god of the earth. Ireland's Holly King is responsible for winter, his twin, the Oak King, for summer. In Greece, there's Apollo, the sun god, and his twin, Artemis, goddess of the moon. There may be no obvious value hierarchy between the natural-feature twins, but in other tales it's clear which sibling the storytellers favor. Ancient Greek and North American myth twins, for instance, are often produced by their mother's sleeping with different fathers, one mortal, one divine, in quick succession. As a result one twin has superpowers and the other's kind of average. Zeus was involved in several of these "heteropaternal superfecundations," resulting in Hercules on one occasion, Helen of Troy on another. Though semidivine Helen was world-historically compelling, her fully mortal twin, Clytemnestra, is mainly famous for stabbing her husband in the bathtub. Similarly, though Hercules was Hercules, his twin, Iphicles, definitely wasn't. When the boys' evil stepmother threw a couple of vipers into their cradle, Iphicles crawled into a corner and wailed, while his brother grabbed the snakes around the neck, one in each hand, and shook them like rattles.

The most charged contrast isn't between the Strong Twin and the Weak Twin, but between the Good and the Evil Twin. Iroquois mythology gives us Good Mind, a kind of beneficent farmer, who scatters attractive and delicious items over the earth's surface for human use. His twin, Warty One, spends his time unraveling his brother's fine work, dispersing hordes of mosquitoes and studding bushes with thorns.

The Iroquois were careful not to make Warty One hotter than his brother. No such prophylactic existed in Hollywood. For some time now, the Evil/Good Twin divide has mapped pretty neatly onto the Sexy/Frigid Twin divide, especially when the twins are female. This trope is rolled out to perfection in a series of midcentury film noirs involving an otherwise identical virgin and whore fighting to death over a man. In *A Stolen Life*, for instance, we meet Bette Davis as Kate, a lonely, passive artist in baggy pants, who falls in love with Bill, a dreamy lighthouse keeper, while staying at her family's vacation cottage on Nantucket. Kate and Bill seem destined for happiness, until Bill runs into Bette Davis as Pat, Kate's serially man-killing twin. Pat knows Kate is crushing on Bill but (or therefore?) decides to seduce him anyway and pulls off the maneuver instantly. (Kate is nice and all, Bill pants goggle-eyed in the garden, but "a man likes a little frosting.") Poor Kate has to endure Pat and Bill's wedding and years of what she assumes, from a tortured distance, is their perfect marriage. Pat, meanwhile, proves superficial and faithless. She continues her erotic escapades unconstrained by her marriage, while Bill abandons his beloved seashore to finance her lavish urban lifestyle.

The instance of this trope that made the biggest impression on Julia and me, growing up in New Zealand suburbia in the 1980s, was Elizabeth and Jessica Wakefield, the identical heroines of Francine Pascal's Sweet Valley High series. Each of the volumes reminded us within three pages that Jessica and Elizabeth were svelte, golden-skinned, blonde Californians, blessed with "eyes the color of the Pacific Ocean." Julia and I lived right by that ocean, but as far as you could get from the Orange County end of it without hitting an ice shelf, and about as far as that, developmentally, from Jessica—especially in the unforgettable scene in *Playing with Fire* where she's in a darkened cabana while her spiritual equivalent, Bruce Patman, plucks insistently at her string bikini. ("Why?" I remember wondering.) Like our friends, Julia and I had almost nothing in common with the Wakefield sisters, but like Jessica and Elizabeth, we were twins, and we gloried in the special status that conferred on us. We pored over the worn pages of the inaugural SVH volume, *Double Love*, like scholars of a sacred text, in search of advice on how to live beautifully. The first order of business was to work out which twin we each were, which didn't take long. Liz (*Lena* starts with *L*) was the introvert: caring, responsible, and sincere, a wearer of cream cashmere, and a self-satisfied, moralistic prude. Jessica, the extrovert, was four minutes younger (exactly like Julia), assertive, status conscious, a sporter of short skirts and distressed denim, and a scheming, narcissistic seducer. As a description of us at the time this wasn't fully accurate, but we had another six years to go before we turned sixteen, so maybe, we figured hopefully, we'd grow into it.

• •

This persistent tendency to binarize twins is striking, especially in light of the exactly opposite companion tendency to portray twins as highly similar. What explains it?

One possibility is that it's a quick and dirty solution to a sometimes panic-inducing problem: that of stably tracking which twin is which. Especially in their younger years, many identical twins are borderline impossible to tell apart. Even the people who produced them struggle to get it right, resorting to external markers, such as painting one infant's fingernails red or attaching a bracelet to one tiny wrist or ankle. For the first few years of elementary school, Julia and I wore our initials pinned to our clothes: an *H* and *J* made of cream plastic or dark coconut shell, whichever would stand out best against the day's dress or sweater. Later things got easier, after Julia lost a front baby tooth in a bumper car incident, and, later still, after I started wearing my hair shorter than hers, she became taller, and our clothing choices drifted in different directions. Still, for some observers, the difficulty in distinguishing us persisted well into our twenties.

Once, when I was back from the United States visiting our hometown in my early years of graduate school, a man I'd long nursed a silent crush on dashed up to me, clasped me to his chest, and planted a kiss on my cheek; it turned out Julia "Jessica Wakefield" de Bres had slept with him the previous week. This kiss happened on a street where Julia and I had worked in our teens, in different branches of the same bookstore chain. Occasionally when we didn't have a sought-after book at hand, we'd send a customer down to the other branch. "How did you get here faster than me?!" they'd gasp. It's hard for us twins to empathize with such people in our own case, since we find our physical differences obvious, but Julia and I once accidentally simulated the phenomenon in a bathroom at our high school. Unknown to each other, we were in adjacent stalls and flung the doors open simultaneously. There in the mirror in front of us were two apparently identical fifteen-year-olds in identical school uniforms. It was super creepy; we screamed.

It's embarrassing when you can't identify the person in front of you, or when you attempt to do it and mess it up. But twin identity confusion also unsettles us for deeper reasons. It's easy to forget, since we don't have cause to think about it often, that the ability to reliably and accurately assign a unique identity to those we meet is an absolutely basic requirement of social interaction. Human societies depend on cooperation, and cooperation requires that we keep track of who's who: who has which tendencies, talents, information, and roles. Cooperation also can't be sustained over the long term without the ability to reward and punish those who comply with or break social rules—and to do that you need to be able to work out whether the person you're dealing with now is the same person you were dealing with last week. The social threat twins pose in this department is dramatized in the 1946 film *The Dark Mirror*, which stars Olivia de Havilland as Terry and Ruth, a pair of identical twins who are both suspects for a murder because no one can tell them apart. As the detective on the case notes, whichever of the two has knocked off the deceased has, by virtue of her identical twinhood, committed "the perfect crime."

Because social cooperation is essential to our well-being, we have self-interested reasons to want to stably track one another's identities. We have ethical reasons, too, because a large chunk of our moral lives depends on reliable identity detection. We can't hold people responsible for their past actions, demand they keep their promises, claim our entitlements from them, fulfill our duties toward them, or compensate them for their sacrifices if we're not sure who they are.

When our identity detector hiccups, as in the case of closely similar twins, the people in front of us therefore slip, at least temporarily, outside the social and moral world. We can neither reliably predict their actions nor hold them to account for their behavior afterward. We're unnerved, and for good reason. "You're a difficult problem . . . ," says K to a pair of twins in Kafka's *The Castle*, "you're as alike as snakes."

The many tales, in real life and art, of twins intentionally switching places with each other or being innocently misidentified reflect our uneasy fascination with the possibilities here. Often twin-switcheroo plots are comic, as in Shakespeare's twin plays, *The Comedy of Errors* and *Twelfth Night*. The first stars two sets of separated identical twins, arranged in matching master-servant pairs, who turn up in the same city and inadvertently cause havoc. The second play dials it down by employing only one pair of different-sex twins, who are alike enough that one can pass for the other in drag. Believing that her twin, Sebastian, has drowned in a shipwreck, Viola adopts his gender identity and offers herself as a servant to the dashing Duke Orsino. In acting as intermediary for her employer in his doomed suit for the lady Olivia, Viola unintentionally serves as wingman for her twin instead, when Olivia falls for the cross-dressing Viola, and then for Sebastian, when he turns up alive just in time to salvage Olivia's straightness. (Whatever you need to tell yourself, Olivia!)

Even when comic, tales of twins switching places tend to have a subversive edge. Twelfth Night refers to the festival of the Epiphany, twelve nights after Christmas Day: a carnivalesque occasion for cross-dressing, role reversing, and licensed chaos. Twins are a natural fit for this party, given the widespread association of multiple births with disruption of the natural and social order. Mark Twain's novel *The Prince and the Pauper* (1881) picked up the theme a few centuries after Shakespeare with a pair of boys who, though unrelated, share a birthday and look identical. Tom Canty, the son of a poverty-stricken alcoholic, switches places for a day with Edward Tudor, the Prince of Wales, who is about to become King Edward VI. "My idea," Twain commented, "is to afford a realizing sense of the exceeding severity of the laws of that day by inflicting some of their penalties upon the King himself and allowing him a chance to see the rest of them applied to others." A Nigerian hymn accords the ability to advance this revolutionary message not just to those who write about physically indistinguishable characters, but directly to twins themselves, who allegedly possess supernatural powers of transfiguration:

> Twin sees the rich, passes them by,
> Twin loves persons in rags,
> Twin will transform a person in rags into
>   a paragon of
> Royal dress and richness.

• •

We identical twins, then, are tricky, disruptive, even seditious creatures. We *are* the perfect crime. Most people only run into us occasionally, but the experience of doing so, or the simple idea of twins, can enflame broader anxieties about the fragility of everyone's capacity to stably identify anyone. The human recognition module can, after all, spectacularly break down. In the Capgras delusion, the most famous of a set of neurological disorders known as delusional misidentification syndrome, a person believes that a relative or friend has been replaced by an identical-looking substitute. In the related Fregoli delusion, the imagined impostor is a shape-shifter, appearing sequentially in disguise as a variety of the patient's relatives and associates. The sufferer acknowledges that the hosts have distinct physical appearances, but insists they're nonetheless psychologically identical.

In both delusions, the sufferer's friends or family members look like they always have done, but are believed to be radically different from how they once were. We can see this as an intensification of an experience any of us might have: that of seeing a loved one change radically, become psychologically unrecognizable. The result can be deeply disconcerting and make the observer doubt their own sanity. The TV show *Breaking Bad* dramatizes this experience, as Walter White secretly transforms from a repressed chemistry teacher into a murderous drug lord. His wife becomes increasingly confused and distraught, and Walter, to his credit, empathizes. "If you don't know who I am," he advises her, "then maybe your best course would be to tread lightly."

The inability to distinguish twins from each other may also trigger a still more personal anxiety in singletons. If some people out there have a fuzzy social identity, a singleton may think, might not *I* myself be that way? This worry has psychopathological correlates, too. The delusion of "subjective doubles" is the belief that one has a doppelgänger, a being strikingly similar to oneself living an independent existence that may occasionally intersect with one's own. The delusion of "clonal pluralization of the self" is the belief that the world contains *multiple* copies of oneself that are both physically and psychologically identical.

If the literary history of twins switching places is mainly comic, the literary history of doppelgängers is absolutely not. Nineteenth-century fiction is rife with Gothic tales of men haunted and tormented by their shadowy and sinister doubles, usually to the point of insanity or death. Edgar Allan Poe's story "William Wilson" hits all the key notes. The titular narrator describes meeting a boy at his school who shares his name, height, gait, dress, and birthday (Poe's own), though he differs notably in speaking only in a whisper. The narrator is irritated and creeped out by his double, who acts chummy with him and has the insulting habit of insinuating helpful but unwelcome advice. Still, "I could not bring myself to hate him altogether," the narrator reports. The two Wilsons have a lot in common, after all, and they end up becoming "the most inseparable of companions," despite fighting constantly. Ultimately, in a panic, the narrator breaks into his frenemy's room at night and, seeing that even the boy's face is now identical to his, flees the school in horror. The double then trails him for years, like Victor Frankenstein's monster, and exhibits a special talent for turning up disapprovingly just when the increasingly corrupt narrator is committing an immoral act. In the final scene William drags his double into an antechamber at a party, engages him unwillingly in a sword fight, and stabs him to death. But whose face is it that then appears pale and bloodied in the mirror, whispering no longer? "You have conquered, and I yield," it intones, in the final paragraph of the story. "Yet henceforward art thou also dead—dead to the World, to Heaven and to Hope! In me didst thou exist—and, in my death, see by this image, which is thine own, how utterly thou has murdered thyself." In this and other classic doppelgänger tales, the double functions as both the Good Twin—morally and socially superior to the original; and the Evil Twin—supernaturally sinister and a harbinger of dissolution.

Most of us these days are inclined to dismiss doppelgängers as a merely literary or psychiatric phenomenon irrelevant to our everyday lives. Maybe we're wrong about that. Over the past decade a series of websites, including twinstrangers.net and iLookLikeYou.com, have allowed you, with the simple upload of a photo, to add your face to a global database where it can be matched via facial recognition software to people who closely resemble you. The allegedly millions of people who've done this so far clearly haven't spent much time reading nineteenth-century Romantic literature. Sure, it might be fun, initially, to find what one website claims are the on average seven total strangers in the world who look exactly like you. (Irishwoman Niamh Geaney has found three of hers so far: Karen Branigan in Dublin, Luisa Guizzardi in Genoa, and Irene Adams in Sligo, all of whom resemble her more than my actual twin now resembles me.) But as William Wilson would warn us, you can't be sure of getting a grip on these elusive characters, and as a result of them, you can't be sure that anyone else can get a stable grip on you. They're out there wearing your face, doing and saying anything they please, completely beyond your control. If they wanted to, they could steal your identity, and in an instant your whole world would come tumbling down.

• •

One way to see the widespread tendency to binarize twins, then, is as a panic response, a knee-jerk defense to the social, moral, and existential threat they pose. Twins remind us, consciously or not, of how frail human identity-detectors can be, and therefore how slippery our associates and our own selves might be. Tagging twins as binary opposites is a way of corralling their disturbingly similar bodies and minds into easily distinguishable ends of the psychosocial field. The lack of subtlety is the point.

Twins aren't the only objects of binary thinking, though, so likely something more general is going on, too. The psychotherapists Jack Denfeld Wood and Gianpiero Petriglieri argue that the disposition to binarize is pervasive in our species. We exhibit an "inherent dualistic psychological pattern" that reduces "complex phenomena or choices to a binary set of alternatives." We cast what might more accurately be seen as variations on a common theme as mutually exclusive opposites, magnifying and solidifying their differences and ignoring their similarities.

Wood and Petriglieri locate the roots of this tendency in the amygdalae, a pair of almond-shaped neural clusters deep in our brains that respond to environmental stimuli by producing intense emotions such as attraction, fear, anxiety, and sadness. Amygdalae help us detect threats and react swiftly and decisively to them, bypassing the more precise but slower cognitive processes of the cerebral cortex. "The amygdala decides whether we like the object or not (and often initiates a behavioral response) before the cortex has even managed to figure out what the object actually is, and long before we are allowed the luxury of a conscious thought or conscious feeling." In effect, Wood and Petriglieri suggest, our brains are primed to instinctively divide the world into "good" and "bad" parcels, and what conscious thinking we do afterward is strongly driven by this initial, broad-brush assessment. If our brains are inclined to be binary about goodness and badness, it wouldn't be surprising if the general tendency extended from there to other matters, too.

An evolutionary explanation for this habit of drawing coarse distinctions is that it has major survival advantages. Those of our ancestors who thought, "Well, it looks like a lion, and lions are usually bad, but it could be a good lion, or a neutral lion, or maybe just a lion-shaped sunbeam," died. Ideally, the amygdalae make the first binary pass, and the cerebral cortex refines and corrects, adding increasingly finer distinctions, the two neural complexes exchanging information and working together to fuse emotion and logic toward reality-tracking action. Sometimes, though, the initial coarse binary sticks.

That's a scientific take; you can also approach the virtues of binarization more philosophically. A core thesis of the twentieth-century intellectual movement known as structuralism is that all meaning depends on contrast. You don't know what something is unless you know what it isn't: how it relates to other things, what broader structure it fits into. If that's so, opposing the object of your interest to some other thing is the absolutely basic move when trying to get a grip on it.

Hegel argued, more ambitiously, that all of history is driven by the meeting of a *thesis* with its tension-generating *antithesis*, and their eventual overcoming via a *synthesis*, which becomes the next thesis; rinse and repeat. The result, he claimed, is a gradual ratcheting up of consciousness toward ever-greater self-realization. That's a dramatic way of putting it, but the general idea that intellectual progress requires a two-step movement, a thought and counterthought, is super plausible. Researchers who study the way creative people think have found that they're drawn to divergence, the meeting of opposites. They call it Janusian thinking: the capacity to hold, as F. Scott Fitzgerald put it, "two opposing ideas in mind at the same time and still retain the ability to function." You don't need to take on all of Hegelian metaphysics to recognize the ubiquity of the dialectical method in philosophy, which proceeds via a repeated back-and-forth movement between an argument and real or imagined objections to it, resulting in an increasingly well-defined and well-defended thesis.

• •

Those all strike me as good explanations for why singletons binarize twins, and they go some way toward explaining why we twins binarize ourselves as well. Just as singletons have practical reasons to want to stably identify us, we have practical reasons to want to be stably identified by them. We can't get what we desire and deserve from others if they don't know who we are. We also have just as much reason as singletons to find the general slipperiness of social identity unsettling, and therefore to want to pin it down rapidly in our own case. And we're humans, with the binary habit arguably built into the human brain.

But, when I reflect on my own experience, I don't think these are the main reasons why twins binarize themselves—or, at least, why *I* did it from a young age.

My junior self was always eager to work out *which one she was*, to pick herself out in the set of two-option person menus she detected all around her. When I was a kid, pairs of fictional characters didn't need to be actual twins, like the Sweet Valley High sisters, for me to pull this move. Whenever a duo came tripping along in a book or TV show, I knew instantly which one was me and which was Julia. I went for the cautious, anxious, or melancholy types, the ones who trotted after the sunny or manic hero, providing assistance and advice and raising useful objections. In *The Wind in the Willows*, I was Mole to Julia's Ratty; in *The House at Pooh Corner*, Piglet to Julia's Pooh. Julia got Mickey Mouse, I got Donald Duck; I got Bert, she got Ernie. In other stories, or in pictures or music, inanimate objects and abstractions could do the work: gold and silver, day and night, major and minor chord, sand and sea.

Even now, as a grown adult, whenever I come across a pair of twins, I feel my mind set off on that instinctual search. I'm looking for my twin among the twins, the one who shares my soul, and also for my twin's twin: the Julia character, the one who'd be my special friend. The answer is usually obvious to me straightaway. Last year, for instance, I read a *New Yorker* profile of the twin poets Matthew and Michael Dickman. Matthew's work, we're told, "is full of tender images of his brother": it "evokes a love forged before the dawning of consciousness, and expresses a striving to replicate that merging in every subsequent encounter—with an erotic partner, or with entire populations." Michael's poetry, on the other hand, "speaks of the ultimate solitariness of the individual, twin or not." The psychologist René Zazzo suggests it's always like this: one twin, he says, is in effect Minister for the Exterior and the other Minister for the Interior. Whether or not that's true, there Julia and I were on the page: Julia rushing toward all of humanity, with feeling spilling out of her chest; me activating the dam for everyone's safety, preparing my silent retreat. I liked Matthew better—God, who wouldn't?—but I was Michael, no doubt about that.

I don't know how the Dickmans felt about being so obviously binarized in their profile. Despite my own tendency to do it, I always assume it's a sensitive subject with twins, to be approached delicately, if at all. Of course, Julia doesn't have that kind of compunction.

We recently met an identical twin we were both eager to impress. We knew a bit about her and her sister, but not enough to perform our usual deduction. They were both writers, they both seemed intense and a little retracted. But they couldn't both be me. As we talked to this woman, my mind wandered to and from the question. Then Julia just straight up *asked* it.

"So one twin's usually the quiet one and one's the loud one. We've been wondering"—we had of course, but, Jesus, we hadn't *discussed* it!—"which one you are?"

"The quiet one," the twin answered in a microsecond.

I was both mortified and grateful, a state of mind Julia is almost uniquely able to produce in me. *Phew*, I sensed the two of us thinking, *now we know which one of us we're dealing with*.

Probably not all twins have this "Which one am I?" tendency. But I bet many of them do, because twins have special reasons to binarize themselves, beyond the reasons that apply to everyone. One is that singletons are always doing it to us in particular, and we internalize the move. In many cases, the process starts at home. Parents get the binary characterization of their twins rolling, for their own reasons, and then their babies, natural learners and people pleasers, try to live up to the pictures of themselves they detect in their parents' behavior. A positive feedback loop ensues, as the infants' mirroring reinforces their parents' initial classification, which gets retransmitted all the more strongly, further encouraging the twins to embody it. Then everyone else in the vicinity catches on and we're all off to the races.

As they get older, twins are also likely to internalize a set of twin binarizations that originate outside their intimate sphere. The usefulness of binary thinking in general as an intellectual tool means that twins, a natural hook for it, are often dragooned as avatars into broader political, cultural, and philosophical debates that have nothing to do with twinhood per se. Twins in myth and art are usually not simply characters of interest in their own right, but symbolic representations of two opposing values, principles, or forces. They're used in the story to clarify the nature and relationship of the things they represent, and sometimes to argue for the social victory of one over the other.

Myths involving nonidentical male twins who found cities or nations are often like this. In Genesis, Jacob, one of the three patriarchs of Israel, is a smooth-skinned herdsman of ambition, steadiness, and intelligence, while his brother Esau is a hirsute hunter who tends toward rash, shortsighted decisions and obliviousness. Jacob uses his native forethought and cunning to cheat his firstborn brother out of his superior inheritance not just once, but twice—first in a well-timed bargain involving a bowl of stew, when Esau returns from work starving; second, by donning a hair suit and bearing fake venison, so their half-blind dying father mistakes him for Esau when giving the crucial final blessing. The moral seems to be that a little deceit is forgivable, even desirable, in someone who has the talent and will to lead a nation to heaven.

When the twins are female, the trope has traditionally been used to explore competing norms of femininity, as in the virgin/whore films I mentioned earlier. On a less political level, twins can function narratively as models of available selves that each individual might privately pursue: representations of contrasting, perhaps incompatible modes of living.

These artful uses of fictional twins recur in everyday responses to nonfictional twins. When confronted with us, people feel the need to examine and categorize us and, often, to take sides. We become real-life characters in a set of two-person morality plays staged for our audience's personal development or philosophical enlightenment. When Julia and I were growing up, some of our friends and relations felt the need to explicitly announce which of us they'd identified as being closest to their soul, as if we were options in a personality questionnaire. With the kids at school and our cousins, the answers were erratic: one day I'd be Victoria's or Jenny's soulmate; by the weekend Julia would have taken my place. Adults were more consistent, but not much subtler, and nor could they be guaranteed to get it right. I was out for a walk once with a woman of such extraordinary loudness and gregariousness that she often frightened me. "I'm glad we can get some time alone," she said conspiratorially. "I love Julia, but I have so much more in common with you."

• •

I recognize all of this at work in my own case, but—you know what?—I still don't feel it gets at the core of why I binarize Julia and me. The center of my experience of the phenomenon doesn't seem to lie in the intellectual functions twin binarization performs for singletons, which we twins then internalize. Instead it lies in the emotional needs binary characterization directly serves for twins. Much of what I'll say here extends to other siblings or close couples, so non-twins will recognize themselves in it, too, but twinhood tends to bring it to a higher pitch.

One thing binarization does for twins emotionally I've already touched on: it allows us to pin down our identities, to angle in on who we are. The self likely appears to a developing human as just as much of a "blooming, buzzing confusion" as the external world, in William James's famous phrase. That sensation is alarming, as anyone who's been through an identity crisis later in life knows. Even if the world appears quasi-stable, how are you supposed to function in it if you don't know where your boundaries are, what holds you together over time, what defines you as you?

The classic primal move is to do it by a process of elimination, starting with your mother. *Who am I?* the infant begins to foggily ask. *Well, okay, not her!* But with twins raised together, a second version of this move has to happen, too. *Wait, not her either!* It's a broad-brush assessment at first, and then, as the twins age, they begin to pin down the details, sorting themselves into two separate camps. The twin psychologist Ricardo Ainslie suggests that radical dichotomizing is valuable for growing twins "in that it becomes the mooring of a 'self-system,' and it is around these distinguishing characteristics that each twin's sense of self as separate from the co-twin becomes organized." The result may be reductive, and cramping in certain ways, but if the prize is existential security, we'll take it?

Beyond that benefit, I think binarization gave me a feeling of power as a kid, a sense of where my talents and resources lay. As well as the introvert, I identified early on as the writer in the de Bres twinship. I tore through every book I could find in the house—novels, medical manuals, recipe books, the telephone directory—and could think of no better adult life than being an author, scribbling out manuscripts like frenzied Jo March in her attic. Julia's career ambitions were more fluid: first she was going to be a tollbooth operator on the bridge spanning the two shores of Auckland harbor, then she moved through lawyer (the answer to "Which job makes the most money, Mummy?"), actress, ventriloquist, diplomat, and barista, till her adult stopping point at linguistics professor. But one thing she loved throughout was drawing, so sometimes we paired up in a junior Virginia Woolf / Vanessa Bell arrangement. When I was twelve, I wrote a novel set in Victorian England, and Julia did illustrations for it: colored-pencil pictures of my characters in pinafores and crinolines, slammed up against the intricately wallpapered rooms of their mansion (the illustrator was still working on perspective) or running wildly through the fields, looking even more real than I'd seen them in my head. Julia *not* being the writer made me feel more like a writer: it validated and fueled my plan.

I was also, in a more unspoken way, the twin who held it together. The general setup was already in place at our fifth birthday party. At the climax, we were gathered around the dining table, at the center of our crowd of admirers, waiting for the cake to be brought out. It emerged suddenly from the kitchen, aloft and aflame: a sculptural wishing well covered in M&M's, with log-shaped cookies supporting a rustic roof thatched in chocolate shavings. We both instantly registered it as sensational, and the realization threw Julia over the edge. She flung herself under the table and had a full-scale tantrum of joy, while I sat above her smiling benignly at the guests, waiting patiently for her to get it out of our system.

I could be relied on as a kid to accept my fate, to cause no trouble, and to talk Julia off the ledge when she reliably failed to do both. Of course, the underlying setup was more complicated than that, and the occasional crack in the façade was visible. Everyone assumed, for instance, that Julia would find the transition to elementary school tougher than I would, but when the teachers tried to place us in separate lines on our first day, I, rather than she, lost it. I remember the panic ripping through my chest when I recognized what was going on, that they were taking Julia from me. The surface content of it was *I have to look after her! She won't be okay without me!* But who was being looked after there, really? Julia might or might not have needed me to look after her; what's clear is that I needed to look after her, and that exercising that power, seeing it succeed, was how I looked after myself. I made a scene that morning, broke character to salvage my role, and then, with my goal achieved, settled like magic into the quiet one again.

Closely related to the sense of power binarization offered me as a kid was the sense of specialness it gave me. Feeling as if you have your own thing, that in your social sphere, small though it may be, you're irreplaceable, is a way of feeling your life is meaningful, that you're making a valuable difference to the universe, that your absence would be a loss. It's natural for identical twins to feel this impulse particularly strongly, given that we sub in for each other easily in so many ways. Drake uses this idea in Rihanna's song "Work" to persuade her to meet him for a hookup: "If you had a twin," he sings, "I would still choose you." Hypothesizing a less desirable twin is one way to make someone feel special; identifying a binary and locating yourself and your actual twin on either end of it is another. *I'm the introvert, I'm the writer, I'm the non-fall-apart-er:* I have a twin, but you could still choose me.

Psychologists have suggested that binarization doesn't only serve the personal needs of individual twins, but also the needs of their twinship. Setting up a division of labor in which each twin has a distinct sphere of expertise and operations allows the pair to minimize conflict in their relationship. If Julia is the illustrator and I'm the writer, we don't need to fight over the same resources, or to compare our talents and achievements directly to each other's. Rather than being in competition, we can be complementary, a setup that has its own aesthetic/mystical satisfactions. ("All things go in pairs," says Ecclesiasticus 42, "one the counterpart of the other; he has made nothing incomplete. One thing supplements the virtues of another.")

Sometimes what's going on here is a mutually beneficial outsourcing: each twin can count on the other twin's special talents for their own use as needed. Other times the division of labor seems to come from a darker place. Dumping your twin in a camp sharply opposed to your own can be a defensive move, a projection of your undesired or suppressed features onto the other. By appearing as your antithesis, in the extreme case as your evil twin, your sibling can soothe your self-doubt and deflect the attention of others from what you don't want them to see in you.

I'm not crying, you're crying. I'm not too much, Julia is.

• •

When I think about all of this for too long, I start to feel nauseous. The somewhat desperate psychic needs binary thinking serves, its obvious disregard for further relevant distinctions, make me doubt my grip on everything it touches, including myself. I've suggested that we humans binarize to make the chaos we encounter easier to manage: to make our surroundings seem more orderly and predictable, and ourselves seem powerful, special, and good. But something's being useful or gratifying doesn't make it true. (God, if only.) When I consider the messy history of my beliefs about what my own self is like, I don't seem to have any reason to trust them. Am I really the quiet one, the together one, the Minister for the Interior, or is that just a story I've come to tell myself for my own comfort? Is my authentic self actually something quite different from the one I detect in operation today: Is some other, truer, primitive Helena lurking in the depths? If so, how could I possibly locate her after a lifetime of misdiagnosing and quashing her? Or—and here the vertigo ramps up another level—maybe there's no *there* there: maybe everything we tell ourselves about ourselves is just a front, and the self is an illusion all the way down.

The dominant intellectual position on this question since the second half of the twentieth century has been that last one. Following Derrida, Barthes, and Foucault, sophisticated people are meant to think it's obvious that the self is a phantom, or that, if it does exist, it's only a performance, an act with nothing of substance backstage. We humans used to think some essential nugget was inside us—a coherent, steady, enduring soul that made sense of all our scattered sensations and survived all the changes of our material body. But that mystical, wishful idea has to be ditched now that God is dead and the sciences have revealed just what a hot mess we all really are.

Call me old-fashioned, but I've never been tempted by the more radical postmodern claims about the fluidity of the self. Sure, we humans are complex, evolving things, with surprises round every corner. But I still think the self has a lot more stability than the average postmodernist claims it has, as I'd suggest a survey of anyone's social circle reveals. The idea that the self is in pervasive eternal flux sounds plausible until you watch your friend, or yourself, dating the same kind of asshole over and over. Similarly, I'll sign up for the idea that our selves are in some way performative: our identities involve a series of suits we try on and switch up as context demands. But I still think some of those suits fit better than others, and that's because a genuine actor is wearing them: it's not just a set of floating costumes filled with air.

Is there a way to insist that the self is a real, persisting thing without sounding naïve? And is there a way to salvage the idea that we know, at least partly, what that real self is like, even while acknowledging the powerful forces of internal and external suggestion on our sense of who we are?

A first step is to note that something can be socially constructed while at the same time being real and having a significant degree of stability over time. Marriages are socially constructed—they're a human invention, dependent on a set of shared beliefs, commitments, and practices, not built into the essential fabric of the universe. But marriages are real all the same: it's a *fact* that spouses are married to each other, and that fact has major practical consequences out in the world, sometimes for decades running.

You could accept that, I guess, and still doubt that the socially constructed *self* in particular is real, semi-stable, and knowable. More to the specific point here, those of us—not only twins—whose sense of ourselves is pervasively affected by binary thinking could still worry that that kind of construction is fundamentally distorting. But what if binary thinking could, at least partly, form our self, rather than throw us off the scent of it? I've read a pair of books recently that have helped me angle in on this question, in ways that make me feel less unmoored.

One is Alexander Nehamas's *On Friendship*. A central argument of that beautiful book is that our selves are to a large degree relational. We don't come to our intimate friendships fully formed and set in stone; instead, "who we are is to a great extent determined by our friends." Part of what it is to be a friend, Nehamas says, is to be open to change and growth in response to the conversations and activities you share. Close friends are willing, in fact eager, to have each other's desires and values shape their own. They try out new ways of being and thinking in each other's company and let each other's responses affect which they continue to pursue and which they discard. Sometimes explicitly—by reflecting and interpreting each other's actions, or by drawing each other's attention to something unnoticed and worthwhile—but often in small, subtle, unarticulated ways, they help each other work out what matters to them, what kind of person they each want to be.

Friendship, by these lights, is an ongoing mutual self-creation, similar to a jazz improvisation (as another philosopher, Ben Bagley, puts it), where the product isn't only a piece of music—the friendship—but the musicians themselves. You come to your early intimate friendships with some preexisting materials, sure, but they're accidental and scattered, picked up here and there from your social environment and history, interacting with your biological makeup. To be a real self, to have an identity that's distinctively yours, you need to find a way to sculpt something out of that random inheritance. Every new friendship has the power to extend our self in new directions, and successful friendships continue to do it lifelong. The self, Nehamas says, is "always unfinished business."

The self is real in this view, but you don't discover it, you create it, and you don't do the creating alone. I find this heartening; if it's true, there's nothing inherently suspect about the sense of self you arrive at by opposing your nature to that of your close friend, or your twin. Doing this can be a useful, respectable way of homing in on what you want to be, on who you will become. You can authentically be what binarizing makes of you.

Many of these same ideas appear in Marya Schechtman's book *The Constitution of Selves*. Schechtman's focus is on the role of narrative, not friendship, in forming who we are. But the same basic picture is there: the self as something not given but crafted, in an ongoing way, in dialogue with the social world. Schechtman argues that we organize our experience of life by telling ourselves, at least implicitly, an autobiographical story. And it's by highlighting certain personality traits, values, actions, and plans as central to our past, present, and future, that we become a person. (Before that, she says, we're just experiencing subjects, human animals with no genuine *self* to speak of.) As in Nehamas's account, this self-construction is a dynamic, open-ended project—as we move through life we continually update our interpretation of ourselves and our trajectory—and we don't do it alone: we draw on others' interpretations. If our story is to do the job it's intended for, it has to fit closely enough the stories others would tell of us. Moreover—though Schechtman doesn't say this, I think she'd agree—we rely on other socially available narratives when crafting our own.

To draw the directly personal moral: maybe it's not inauthentic for your sense of self to derive pretty obviously from a set of parental expectations and binarized fictional characters you were exposed to prior to your teens. Or, if that is inauthentic, maybe the authenticity it's opposed to is neither possible nor desirable for such deeply social beings as us.

• •

The most intractable problem with binarization, even if Nehamas and Schechtman are right, is that it threatens to narrow your sense of the possibilities you contain and trap you there permanently. If the self is a voluminous and ever-evolving thing, if each person is, as Nehamas says, inexhaustible, binary thinking can't hope to accurately capture it at any one moment, let alone over time. We can't step totally outside the self-images our relationships

generate or the self-narratives we spin. But we can look for cracks in them and teach ourselves to tell more complex stories that fit our inner life and outer behavior better. The search for the authentic self isn't a search for the pre-social or asocial self, but for the capacious self, the self that lives up to its multidimensional character, that lets itself surprise itself again and again.

One nuanced alternative to seeing twins as binary opposites is to see them as polarities: each occupying one of two ends of a continuous axis. This picture implies that each twin's characteristics, though different, aren't alien or hostile to the other's. It also implies that those characteristics might change over time, even, in the extreme case, causing the twins to flip positions, coming to inhabit the opposite pole from the one they started in. Claude Lévi-Strauss noticed that Native American traditions favor the polar model. Unlike Indo-European myth, which tends to treat twins as either identical or antithetical, "the New World prefers intermediate forms . . . American myth incorporates disparity and does not seek to amend it." One twin is often given the role of the trickster, a recurring mythic character who represents "an unstable duality," "the principle of imbalance." The trickster works against symmetry, equated with artificiality and stagnation, in favor of disequilibrium, movement, and life. "He knows neither good nor evil yet he is responsible for both. He possesses no values, moral or social . . . yet through his actions all values come into being." He certainly can't be pinned down as the simple and enduring opposite of his twin.

Another possibility is to construe twins as neither binaries nor polarities, but as complements, each possessing distinct features that supplement and enhance the features of the other. The features don't need to be oppositional, and they'll often be more valuable to the duo to the extent that they aren't. Nor does the presence of a feature in one twin imply its total absence in the other: one twin can just have a more fully developed version or tend to employ it more widely and more often.

A more radical alternative to binarization is to admit the obvious: that twins are often quite similar. The Greek gods Phobos and Deimos are a nice example of a pair who have clearly dispensed with binary oppositions. Deimos was in charge of terror and dread, we're told, while Phobos was in charge of fear and panic. Thanks for clearing that up, guys.

In some cases, twins are used narratively precisely to undermine binary thinking. A stark, immutable opposition between the twin characters is set up and then progressively complicated, the differences between them moving from a rigid duality toward a more fluid, interesting, and realistic relationship. The implication is that not only twins, but all humans, and possibly all of reality, have this complex and mutable character.

Take the nonidentical twins Aron and Cal in John Steinbeck's *East of Eden*, which relates the history of a family seemingly doomed to intergenerationally repeat the biblical story of Cain and Abel. While Aron is fair-haired, beautiful, sweetly sensitive, and universally loved, Cal is dark-haired, unattractive, awkward, jealous, and violent. Cal's development at first tracks that of his dark uncle Charles and evil mother Cathy (the Cain team), while Aron (like his doting father, Adam, on the A team) remains innocent and trusting. But by the end of the book our sympathies have decisively shifted to Cal, despite his indirectly causing Aron's death on the battlefields of World War I. Of the two, Cal has become the more morally motivated, in his sincere battle to escape what he sees as his evil inheritance, whereas Aron has proven to be so rigidly moralistic, and at the same time so superficial and spineless, that he's incapable of genuine intimacy and growth. In their story, the twins move from binary opposites to something more like polarities. Their role in the novel is to illustrate the possibility that all of us might escape the confines of our original character and circumstances and reach freely toward either good or evil.

The pop version of this plot turns up in Bette Davis's second twin movie, *Dead Ringer*. The film at first seems to follow the familiar good-and-evil-twins-fighting-over-a-man shtick. When we meet hardworking, cash-strapped Edith, she's estranged from leisured and loaded Margaret, who deceitfully stole Edith's wealthy boyfriend some years ago. The twins run into each other when Edith turns up at the man's funeral, unleashing a series of events that, to understate it, make it hard to see Edith as the virtuous party. Edith murders her sister on their birthday, as a belated act of romantic revenge, and moves into Margaret's opulent mansion in West L.A. But just as we're slotting Edith into the evil-twin role, our expectations are overturned once more, when Edith discovers that Margaret's husband died not of natural causes, but at the hands of his wife and her lover. It turns out that impersonating your victim only gets you out of trouble if the victim you're impersonating isn't a murderess, too.

Bette Davis can be read in this film as undoing the standard binarized female-twin trope she'd exemplified eighteen years earlier in *A Stolen Life*. But, examined more closely, that earlier film is itself more subversive than it first appears. In the final third, long-suffering Kate and flirtatious Pat get caught in a storm while sailing, and Pat is thrown out of the boat to a watery death. When Pat's wedding ring slides off in her twin's hand as Kate desperately tries to save her, Kate slips it on for safekeeping. Dazed Kate is then mistaken for Pat, and passive as ever, but now in a rather more interesting way, Kate sort of, kind of, absolutely lets the misconception continue. She gives it up eventually, but not before she's moved into her twin's house, discovered her sister's infidelity, and fallen into Bill's grateful, repenting arms. If Kate is the good twin, why exactly is she kissing her recently deceased twin's husband just after committing a felony?

. .

Part of what I like about these more complex twin stories is that, while they put serious pressure on binary thinking, they don't fully dispense with it. If the binary habit comes with being human, we can't get rid of it entirely, and, given its uses, maybe we wouldn't want to even if we could. But we can *mess* with it.

One way to mess with it is to recognize yourself in your alleged opposite. I found this easier to do when I got some geographical distance from Julia, after moving to America for graduate school. It was the first time I'd spent any extended time away from my twin and was almost the first time I'd faced a new social scene without her at my side. I pretty quickly discovered two things. First, I wasn't as introverted as I'd thought. Cambridge, Massachusetts, is stacked with pro-level introverts, holed up in libraries and labs, avoiding one another's eyes in the grocery aisles, calling it a night at nine thirty to retreat to their books, tea, and cats. As I (inchingly) befriended some of these people, it became clear to me that, while I may be reserved and low-key compared to Julia, compared to others I'm definitely not. Or at least I'm unusually good at faking Julia's side of the equation in public. "Professor de Bres is wonderfully happy all the time!" one of my early teaching evaluations proclaimed. On reading this, I moved from *wtf* to *lol* to *huh*, which is where I've more or less stayed.

My second, related, discovery was that Julia and I were much more similar than I'd come to think. No one was there in our twenties to compare, contrast, and magnify the differences between us, and we lost the daily opportunity to do it to ourselves. As a result, the lines we'd always drawn between our personalities came to seem less well defined. Now when I go home to New Zealand, I love reminding myself how similar Julia and I are. "Everything I am not is walking across the room towards me," says a Michael Dickman poem, "and looks just like me."

Another classic way to liven things up in your binarized couple is to introduce a third. Julia and I sometimes used to imagine as kids that we had a long-lost triplet named Caroline. We didn't wish she actually existed, because we had the standard twin view about triplets, quads, and other "supertwins," which is that they take a good idea and overdo it. But we found the idea of Caroline funny (and a little creepy: she was our tripleganger, after all). The problem with the game was that I couldn't lean very far into it because I had major trouble imagining what Caroline would be like. Was she a Hegelian synthesis, some kind of personality merger of Julia and me? Or was she to the left of one of us in some crucial respect, so that, say, Julia was the synthesis, or I was? But if so, *which* respect of us was she to the left of, and in which order were the three of us standing? Caroline's identity was fuzzy because it was clear that multiple versions of her were available: there was a clonal pluralization of Caroline everywhere I looked.

I now think that what Julia and I were doing with this game was a deep and important thing: we were trying to destabilize the script we'd been telling ourselves since infancy about who we each were. We were doing a little dialectical therapy on ourselves, without a therapist, for free.

Jung is famous for advising us all to identify with our shadow, the supposed Other, our dark and banished twin. "This thing of darkness," Prospero says to Caliban in *The Tempest*, "I acknowledge mine." But what Jung actually said was more complex than that, and harder to pull off. The true route to self-realization, Jung suggested, isn't to integrate your other half into yourself for good, so that you become simply one, rather than two. That's as much a route to inertia as keeping your shadow at a safe distance. Instead the idea is to maintain the two halves in productive tension, to stay, as they say, with the trouble: "The ego keeps its integrity only if it does not identify with one of the opposites, and if it understands how to hold the balance between them." The real challenge, for singletons as much as for twins, is to be not one, nor two, but multiple: to perform your own lifelong switcheroo, be your own self's permanent trickster.

2

HOW MANY OF YOU ARE THERE?

Julia and I were working at a library a couple of years ago, on projects with tight deadlines, when I heard myself say distractedly, "I'm going to the restroom. Do you want me to go for you, too?"

Julia's concentration-glazed eyes lifted from her laptop, quivered briefly, then settled into an expression of derision.

"Oh, wow," I said, blinking. "What the—"

"I think I will empty my own bladder, thank you," Julia suggested.

"Fine." I replied, collecting myself. "You do you."

As I headed to the toilet, I could hear Julia sniggering behind me. I was amused, too, but also unsettled. We'd spent practically all our time together over the past month, during my annual visit to New Zealand, and this was a vivid sign that I'd lost significant grip on where I ended and she began. I try to ward off a creeping sense of interpersonal merger whenever I spend more than five consecutive days with Julia, but clearly the ramparts had collapsed once again, and my resulting assumption that we somehow shared a urinary system suggested my departure would be wrenching.

I was also annoyed at myself for a more intellectual reason. Some rogue part of me had apparently adopted a view about the metaphysics of twinhood that I've spent much of my life resisting. Singletons have a habit of implying that twins aren't fully distinct people, but rather—somehow—a single person spread over two bodies. To cite a random set of examples: Antonio asks of Sebastian and Viola in *Twelfth Night*, "How have you made division of yourself? An apple, cleft in two, is not more twin than these two creatures." The Nuer people of South Sudan don't hold a ceremony when one twin dies because they believe the surviving twin will continue the life of its sibling. The twin protagonists of Michel Tournier's novel *Gemini* are collectively referred to by a single name, Jean-Paul. And the chief surgeon of the Hôpital Bichat in Paris explicitly announced in 1926, as a matter of medical fact, that single-egg twins are two copies of the same person, not genuinely individual human beings.

If asked directly, most people would deny that they consider twins a metaphysical unit, but their behavior often suggests they're inclined in that direction. Especially as children, we twins are given a single present to share, are referred to as "the twins" instead of by our individual names, and are treated as interchangeable in friend groups or by teachers. Keeping track of who we are doesn't seem super important because we're assumed to sub for each other in most social roles. And people get an evident satisfaction out of grouping us together. When I post photos of me with Julia on my trips back home, they blow up with likes in seconds. Our friends are happy we're having fun, sure. But it's also as if our geographical reunion has mended a puzzling tear in the fabric of the universe and everyone feels better now.

You might think only a singleton would be tempted by the idea that twins are a single or duplicated person, or a split self, but some twins seem drawn to it, too. One developmental study describes a pair of twins who, between the ages of two and four, routinely used a single name, singular verbs, and the pronoun *me* when referring to themselves collectively. After observing them chatting at thirty-six months, the researcher recorded in his notes, "They talked as if they considered themselves one." As for adults, many members of the Facebook group for identical twins that I lurk on refer to their siblings as "my other half," in a way that reads to me as only surface ironic.

Julia and I have distinct personalities; we live independent lives in different countries; I don't have access to her calendar, let alone her thoughts; when someone steps on her foot, I don't *feel* it. If there's any basis for thinking we're one person, I've always assumed, it must be some incoherent or mystical conception of personhood that it'd be not only unprofitable but uncharitable to examine.

That's been my public take on the matter. Then I find myself offering to pee on behalf of my twin. It could have been nothing but a meaningless linguistic slip, but to me at the time it felt more like a trapdoor opening over an inner existential cavern. Whatever my level of distraction, I'm pretty sure there's no one on the planet other than Julia whose bladder I would offer to empty—setting aside myself.

"So some buried part of me believes I'm somehow merged with Julia," I marveled to myself as I checked my (our?!) reflection in the bathroom mirror. "What kind of confused and offensive reasons could I have for thinking *that*?"

Then my philosophy habit kicked in. I shouldn't just assume my reasons were confused and offensive, I told myself, because, you know, maybe they weren't. Could something in the neighborhood of the idea that twins share personhood actually be *right*?

I was interested in the answer for my private purposes because I generally like to know which parts of the universe I'm taking up at any one time, thank you very much. But the answer struck me as having wider significance, too. If the boundaries between people turn out to be messier than our standard picture of humanity assumes, what would that do to each of our understandings of ourselves, one another, and the entire social world?

. .

A standard assumption of modern Western culture, so basic as to generally go unstated, is that each person is physically discrete, cleanly distinguished from all other people by their location, solo, within an unbroken continuum of skin. So one way for twins to assure others, and themselves, that they and their twin are separate people is to appeal to this assumption. "As a quite general matter," I can intone, "bodies correlate one to one with persons. If there's one body, there's one person, and if there are two bodies, there are two people. Julia has a bladder, I have a different bladder, and each is contained within a distinct human-shaped package of skin. Different skin-bounded bladders, different people, people!" This is getting gross, but it does seem to settle the issue.

There's a problem here, though. Conjoined twins, by definition, share a single body, so this line of thought implies they're a single person. In the case of parasitic twins—where one twin is much smaller, incompletely formed, and lacking a head—this might be right. But in the nonparasitic cases, conjoined twins overwhelmingly consider themselves to be two unique beings, and those who meet them agree. Alice Dreger, who's extensively studied conjoined twins, reports that it's typical for such twins to speak of themselves as individuals, in the first-person singular, and to develop a personality and tastes different from their twin's. Their families and friends, too, think of them as two people who just happen to be physically attached.

If that's so, it looks as if our standard assumption that bodies correlate one to one with people is wrong. And recognizing that has potentially large implications. If one body can contain two people, why couldn't one person be spread across two bodies? Non-conjoined twins are the obvious candidates here. But once we twins have broken the body barrier, what's stopping singletons from doing it, too? What makes you so sure that all of you is contained within that single envelope of skin?

• •

Julia and I are "mirror image" twins. I'm left-handed, she's right-handed, my hair whorl is on the left, hers on the right, and we have matching moles, one on my left thigh, one on her right. When we told people this as kids, they'd reply, "Oh, so you're not identical, then?" We considered nonidenticals sad poseurs, so we took this as an insult. "Oh, no," we'd retort proudly, "we're actually *more* identical than normal identical twins." When people replied, understandably, "What?" we rolled out our go-to elaboration. Our egg, we announced, had split later than the egg of standard identical twins. It had got impressively far along in forming a normal two-hemisphere person, with one left side and one right. Then, suddenly, it had split in half, so that Julia got all the stuff on the right and I got all the stuff on the left. "If it had tried to split just a little bit later," we'd add ominously, "we'd have been conjoined twins." Our audience would gasp, their attempt to belittle us vanquished, and we'd trip off in satisfaction.

If by *identical* you mean "exactly alike," no twins are identical, and mirror image twins certainly don't count as especially so. But if by *identical* you mean "the same as," this explanation Julia and I used to give makes more sense. What we were saying, in effect, is that we'd come closer to being the same person than standard-issue single-egg twins had. We'd almost made it to singletonhood, but had dropped out of the game at the semifinal uterine stage, by belatedly deciding not to keep it together.

Conjoined twins stay in the game a little longer, abandoning singletonhood only at the very last hurdle. As a kid, I associated them with a thrilling rarity that I was happy to assert some kinship with. It was cool to have been *almost* a conjoined twin—I mean, how many people can say that? But it was also essential to assert Julia's and my distance from conjoined twins, to point out that we'd escaped their fate. While they were enviably attention-grabbing, their bodies were weird: abnormally shaped, physically constrained, functionally impaired, frightening. Conjoined twins were *disabled*. They were not like us.

This take on the matter glossed over the fact that Julia's and my bodies were weird, too. One reason for that was that we were twins. Non-conjoined identical twins may not share the same body, but they still strike external observers as an anatomical exception, an uncanny doubling of forms that borders on the alarming. Two bodies that are somehow less than two. *Shiver.* Think of the pure horror of those immobile sisters in the hallway in *The Shining.*

But Julia's and my specific twin bodies were unusual for another reason as well. We were born with a connective tissue disorder known as osteogenesis imperfecta (OI), the result of a genetic mutation in our shared egg. The key symptom is a tendency to fracture bones after minor impact or strain. Another is a general weakness in muscles, tendons, and ligaments, which makes spraining things easy and lifting and lugging things hard. A related set of irregularities includes curvature of the spine, tilting of the rib cage, skin that bruises dark and fast, and eyes like miniature robin's eggs, blue where they're meant to be white. And in your later years, pain all the time, as your system, ill-equipped for life, begins to slowly collapse ahead of the normal human schedule.

As kids, our main problems were broken ankles, wrists, arms, legs, and shoulders, of which we had about twenty each before finishing high school. We broke them by tripping on vacuum cleaner cords, slipping on wet marble, tumbling down stairs and escalators, falling off swings, turning on faucets, colliding with lane swimmers, and once, in true New Zealand style, being rammed by a sheep. We got semi-used to the recurring drama of our breaks—accident, X-ray, casting up, six weeks of sling or crutches, physical therapy, then start all over—to the point that they were mainly frustrating rather than traumatic. We thought ourselves lucky to have the mildest of the nine types of OI and to live more or less normal lives between our fractures. Yet we had a background understanding that our situation could, either suddenly or slowly, get a whole lot worse. If we fell from a great height or were punched in the face or hit by a car, the likelihood of major brain or spinal injuries or death was unusually high. Even if we avoided such events, the course our condition would take was radically unclear. Maybe our skulls would gradually sink into our necks, causing neurological damage, or our deformed chests would restrict our lungs, till we couldn't breathe. Maybe our spinal cords would be progressively crushed and our legs would cease to work. Our condition was rare and known to play out in very different ways. Who knew what our version would be? In the meantime, we had to exercise special caution whenever we, or anything around us, moved.

In elementary school, we gave an annual presentation in front of the class to inform any new arrivals that we were fragile goods. As we spoke, the kids stared up at us from the mat, agog and entranced. We looked just like them, but we were made of *glass* and there were *two* of us. Our bare existence blew their little minds dry.

• •

Assuming we take conjoined twins at their word and see them as separate people, what makes them so? A natural answer is that, in the two-headed cases, the twins each have their own brain. If you have two or more separate brains in front of you, this idea goes, you have two or more separate people in front of you, regardless of how many bodies are present.

But this doesn't seem quite right. I recently read about Katia and Tatiana Hogan, a pair of teenage Canadian twins conjoined at the head. They share a thalamic bridge, which connects the thalamus (a structure above the brain stem) of one to the thalamus of the other. As a result they share a single continuous brain, and, at least partly, a site of consciousness. While they each control distinct limbs, some of their senses operate in tandem. As babies, when one was tickled, the other jumped, and when one was given a pacifier, the other was soothed. When Katia puts something in her mouth, Tatiana tastes it, too, and vice versa. Most trippy of all, they can both see out of each other's eyes.

The count-the-brains strategy for determining personhood would imply that Katia and Tatiana are, if not one person, at most one and a half. But that conclusion is hard to credit as soon as you view even small snippets of the sisters on video. Though their conscious experience is partly shared, they have quite distinct personalities: Katia is the expressive, highly strung extrovert, Tatiana the chill and quiet jokester. Doesn't that seem crucial when deciding whether they're one person or two?

The idea we're getting to here is that what fundamentally matters for separate personhood isn't the presence of separate bodies, or separate brains. Instead it's the presence of separate *minds*: two distinctive, relatively stable clusters of beliefs, desires, attitudes, values, character traits, patterns of thought, and tendencies to act. Two beings you could talk to or make friends with, hate or fall in love with. Two separate minds, two separate people—this time, surely, we've cleared it up.

• •

Julia and I were born in late November, which makes us Sagittarians. As a kid, I was proud of our star sign, which was clearly the best one: not a sad-ass water carrier, not a crab, not some goat or sheep or whatever, but a *horse*, or at least the back end of one. Julia and I were pony-obsessed in our preteens and spent what feels like all of the late eighties lying on our living room floor reading our double subscriptions to *Horse Sense* magazine and galloping plastic steeds across the carpet. Sometimes we'd sling two bath mats over the back of the couch and straddle it like a tandem bicycle, waving our imaginary riding crops in the air and shouting "Giddy up!" as the sun set over the prairie—our suburban lawn.

We assumed that real horses were out of the question for skeletal reasons, till our parents heard about an organization called Riding for the Disabled (RDA), which offered equestrian lessons on a farm forty-five minutes from our house. When we got to the corral, we each picked a horse and a team of volunteers hoisted us into position. Then we were led around the ring at the pace of a turtle. You'd think it'd be impossible to fall off a geriatric horse with one sturdy social worker clamped to each side of it pinning your wasted limbs to the saddle, but I managed it once, when my ride broke into a surprise trot and I instantly slid off. Whatever, I wasn't in it for the medal. I spent most of each half hour on horseback absolutely terrified and the rest of the week dying to go back.

I was impressed at that age by the claim that we Sagittarians fell into two camps, corresponding to the two halves of our horse-man. There are the athletic types—those legs and loins and that swishing tail; and the bookish types—that noble head and drawn arrow, aimed forever at the truth. Both types, the zodiac guides claimed, are adventurous. The question is just, do you roam out, or do you roam in?

From a young age, I went firmly for in. I enjoyed reading about young women flinging themselves across moors in a fit of sensual passion, or, especially, galloping over mountain ranges on palominos, but I didn't plan to do any of that myself. I preferred to think about my seriously compromised body only when absolutely necessary, and when I did have to acknowledge it, I didn't identify with what I was contemplating. My body was the thing I used to walk to school, get books off the library shelf, and comb my favorite RDA pony's mane with. I didn't comprise it, I *employed* it. In principle, I could do without it.

It was clear to me, then, which part of the Sagittarian centaur I was located in, and it definitely wasn't the ass end. But I felt affection for the horse part of my sign and a kinship with it that I couldn't quite pin down. It reminds me of how I now feel about serious athletes, who on the face of it are my absolute opposites, except for one interesting feature. The only group of people who have even close to as many fractures as people with osteogenesis imperfecta are those with unusually strong bodies: people who engage in high-impact or high-risk sports, playing football or rugby, flinging themselves down black diamond slopes, scaling mountains, and leaping off cliffs for fun. The physically disabled and the hyperathletic are both testing the limits of the human body, with similarly extreme results; they're both, in a way, mutants. And if you put us nerds and athletes together, in the definitely mutant form of a centaur, we're an oddly alluring monster. What a gorgeous vision, that astral intellectual sliced off at the waist, looking over his shoulder at the twitching flanks behind him. He knows he's a marvel of nature, a mystical conjoined twin, and he mans his restless beast half like a god.

• •

My childhood sense that the core of my person was located in my mind rather than my body changed hardly

at all as I got older. If anything, it intensified, as I enrolled for a college degree, then PhD, in philosophy. Whatever their intellectual views, philosophers aren't known for privileging the physical aspects of existence in their everyday lives. Culturally, we're the archetypal absent-minded professors, shuffling around disheveled, gazing into the middle distance, eschewing sex and personal hygiene and forgetting to eat. We're assumed to see our bodies as encumbrances, their mundane demands annoying diversions from the essential business of abstract contemplation. When Kierkegaard collapsed at a party and people tried to help him up, he allegedly said, "Oh, leave it. Let the maid sweep it up in the morning."

There's a long line of theoretical justification for this type of attitude in the history of Western philosophy. Plato, Aristotle, the Stoics, the Epicureans, and the Cynics all argued that the capacity to reason is the essence of human beings, the part that makes us who we really are. Our rational nature is also the best part of us, the part we share with the gods. Our bodies, in contrast, which the ancients associated with primal urges and passions, we share with the lower animals. It's important to maintain their basic functioning so as not to die, but the truest and best human life requires a significant degree of dissociation from one's physical form. Bodies are messy and unpredictable—tell me about it—and their drives are in fundamental tension with the calmer and more dignified activities of the intellect. Plato compared reason to a charioteer who can't steer straight unless he whips the frothing horses in his employ into significant check.

The ancient Greek picture of persons as essentially disembodied minds housed in unruly bodies was adopted wholesale by Christianity, continued to be popular through the early modern period and the Enlightenment, and persisted in a more fractured and ambivalent way through many of the intellectual movements of the twentieth century. I see its long shadow on my students when the subject of personal identity comes up in class. They're undergraduates, mostly with no formal exposure to philosophy when they first get to me, so they're a window of sorts into the metaphysics of the age. To judge by that window, which may admittedly be narrow, the dominant metaphysics of personhood in Western culture hasn't changed hugely since the first century B.C. Officially, kids these days are hyperembodied. In the self-care, Soul-Cycle, sex-positive present, we're meant to spend our leisure hours applying superexpensive products to our bodies, sculpting them into perfection, and then seducing as many people as possible (at least on social media). The idea that our physical selves are fundamentally unimportant to our identity and value doesn't loom large in that picture. But, if you're looking for it, it doesn't take long to detect a basically Platonic worldview lurking in the undergrowth.

A common thought experiment in the philosophical literature on personal identity asks you to imagine your mind, which is to say all your mental life—including each of your memories, beliefs, intentions, desires, preferences, and character traits—being lifted from your current body and transplanted into a different one. Then you're asked to tell us where you think you are at the end of the operation. If what holds your personhood together over time is your body, if you're anywhere at all, you're right where you were at the outset, lying on the operating table. But if what holds your personhood together over time is your mind, your old body is like the exoskeleton a cicada leaves on a tree, and there you are in your brand-new outfit, tentatively stretching your wings.

Pretty much all of my students go for the second option: when pressed into a philosophical corner, they believe they are where their mind is. Similarly, they believe that if their minds were to be extinguished—for instance, if they fell into a permanent coma—their selves would effectively be extinguished, too, even if their fleshy hearts kept throbbing for decades.

• •

When I tell a friend over drinks that I've begun to wonder if twins are, in some sense or to some degree, the same person, I see her struggling to quickly arrange her features into a neutral expression that doesn't suggest I've lost my mind.

"I mean, obviously me and Julia have separate bodies," I follow up, "but I'm not sure that settles the issue."

My companion looks confused, so I mention the conjoined twins example, then tell her about this Amazonian tribe I've read about. Apparently the Wari people see human bodies as pervasively interconnected, through ongoing exchanges of physical substances such as breast milk, semen, food, and blood. They conclude from this that all people are distributed across multiple bodies, rather than exclusively contained within one.

"So maybe this idea that we're divided up into separate people by the boundaries of the body is just a Western prejudice," I suggest.

"Uh-huh." My companion looks dubious.

"But I'm not really thinking about twins merging across bodies via physical exchanges," I say. "I'm thinking more about them sharing mental stuff—"

"Oh, you mean like telepathy?" she interjects.

Many singletons believe twins can perceive each other's thoughts, feelings, and sensations and transmit information to each other outside the usual sensory channels. Some twins themselves claim to be able to do it, pointing to occasions when they seemed to know things about or via their twin while physically distant from them. Julia experienced something like this when we were in elementary school: it was as if the news that I'd broken my leg on the playground had been dropped into her brain instantly, without her witnessing the event or anyone telling her about it.

Nothing of this nature ever happened in the other direction, which used to make me feel self-absorbed and oblivious, excluded from the more mystical dimensions of the universe, and like a substandard twin. Having done some research since, I feel better about myself. The scientific consensus, after a century and a half of investigation, is summed up succinctly by Michael Shermer as follows: "Twins cannot read each other's minds through ESP or other psychical powers better than anyone else, which is to say they can't do it at all because no one can." The results of experiments allegedly demonstrating telepathy haven't been replicated, and the mechanisms posited by parapsychologists conflict with basic axioms of modern theoretical physics. Plus there are lots of plausible demystifying explanations for apparent instances of mind reading. In the case of close family members or friends, the best explanation usually points to the intimate knowledge those involved have of each other and their backlog of shared experience. This does a good job of explaining why Julia and I were a terrible team to play Pictionary against throughout the nineties. Julia would draw a radish and I would scream, "The north pole!" or "Zsa Zsa Gabor!" We always won.

"Not telepathy, exactly," I tell my friend cagily. I'm not sure where she stands on the matter and I don't want to burst her cosmic bubble. "That's about communicating between minds. I'm thinking more about twins *using* each other's minds—or, you know, using their own mind but outside the skull we normally associate with them."

"Okayyy," my friend says.

"I know that sounds weird," I say. "I mean, maybe it's impossible. But I no longer see an obvious reason to rule it out."

My companion blinks. We're in a bar drinking cocktails and were meant to be having fun. I take pity on us both and change the subject.

• •

Minds depend on brains, and brains are physically contained within skulls. So it's natural to think it's impossible for a mind to extend beyond the boundaries of a single head. But according to philosophers who defend the "extended mind thesis," some of an individual's mental functions do actually occur outside the skull. The original argument appeared in a paper by Andy Clark and David Chalmers that was published the year I declared my philosophy major. I'm told it's become more widely accepted since, but I got the impression at the time that the philosophy faculty at my university considered the idea completely kooky, which was saying something, given the wild shit they otherwise seemed to believe.

Clark and Chalmers point out that a lot of human cognition depends on technological artifacts and resources external to our heads. We couldn't do much of the thinking we do every day without pens and paper, books, smartphones, computers, databases, and (more abstractly) language. We use these things so frequently and pervasively, they're so deeply enmeshed in our ongoing ways of processing the world, that, Clark and Chalmers say, they've come to be an integral part of our cognition. They don't merely aid our thinking, the idea goes, they *are* our thinking, or at least some of it.

This argument relies on the idea that thinking is in essence a computational process that performs a set of functions. If that's true, to identify something as an instance of thought, we simply need to identify a process that plays the functional role that thinking does. It doesn't matter where the process is. If your use of the calculator on your phone plays essentially the same role for you as your tallying up the numbers internally does, we should see both acts as forms of thinking, and provided your phone is deeply and reliably enmeshed in your life, it and your brain should be classed as a single cognitive system.

It's a further step to suggest that someone's mental functions might occur not just outside their own skull, but inside the skull of another human. But if your mind can extend to an inanimate object, why not also an animate person? Clark and Chalmers suggest this is possible in principle, especially in "an unusually interdependent couple." Some empirical work in social psychology supports that idea. Daniel Wegner's studies of what he terms transactive memory explore how couples or groups use each other as repositories of distinct forms of information, allowing each to remember more than they would singly. Couples also "cross-cue" each other, remembering in tandem by throwing prompts back and forth till they trigger each other's recollections, "in a sense," one writer suggests, "Googling *each other.*"

Whether this counts as socially extended cognition depends on how enmeshed the couple's thinking really is. Are their minds reliably available to each other and regularly consulted? Do they trust the information each other possesses to the same degree they trust their own mental states? Does their behavior depend on their joint processing, to the point that their competence would plummet in its absence, much as if part of their brain had been removed? If yes, yes, and yes, we might say that their minds really do extend into each other's head.

I relate to all of this in a big way. For instance, I've realized recently that my autobiographical memory is way worse than Julia's. Whole swathes of my childhood and teenage years feel bleached out or missing: I remember what I was doing generally and on a larger time scale, but except for a few significant scenes or periods, I often have little idea of where I was doing it, who was there, and what exactly happened. This is embarrassing for someone who spends a lot of her time writing personal essays, so I do my best and then call up Julia, either to jog my memory while I'm writing or to fact-check myself afterward. ("Are you sure this is the right literary genre for you?" Julia asks. "I mean, I can see it all in, like, vivid color—" And then I hang up on her.)

I guess any hamstrung memoirist could remember via a sibling or best friend in this way, but identical twins raised in the same home surely have a major advantage. Julia and I did practically everything together till I left the country at twenty-one: we went to the same schools, we lived with our parents through college, we had the same academic interests and leisure activities, we went on the same vacations, and our social scenes were closely entwined. I trust her memories of our distant past as much, if not more, than my own. And when I'm ineptly trying to dredge up the more recalcitrant secrets of my personal history, it doesn't feel all that different from asking Julia to do it instead, except that the second option is way easier.

• •

What I do remember clearly about my childhood is that Julia was always there. When I think of us as kids—at the beach, in the backyard, at school, or at the hospital—I see the scene from my own perspective, and I also see someone very much like my child self present inside the frame. Or I see a tuft of blonde hair at the side of my mental vision and am not sure if I'm recalling my hair or Julia's, in the early years before we went stripy and then dark. For two glorious decades we were each other's constant familiars and adjutants, forever at the ready. I found just about everything Julia did interesting and important, and she signed up for almost anything I suggested. I loved the creative glint we saw in each other's eyes when sharing an idea or plan, the sense of the drums kicking in.

The extended-mind thesis captures something important about the experience of merger I sometimes have with Julia: the sense that I'm thinking via her mind. Yet it doesn't encompass the full range of ways that close twins like us draw on each other's powers, or the deeply social nature of our interaction. Transactive memory, even when reciprocal, is really just one mind picking the other and departing solo. Julia and I went further. We executed our missions jointly, with almost no friction. It was like having an extra jetpack strapped to your will.

Tim and Greg Hildebrandt, twin artists who collaborate on paintings, sometimes work side by side, sometimes in relay, including while the other sleeps. Greg says, "We never wake up to something we don't like. When I awake, I see that more is done. It's like magic—I'm sleeping, yet the painting progresses . . . When Tim paints, it's as if I'm painting." As a result, Tim says, "We sign only our last name to our paintings. It's a joint ownership: it's our painting, not Greg's or mine."

Twins who work this way remind me of the concept of "plural agency" philosophers talk about. There are different ways of spelling it out, but according to Bennett Helm's version, what's crucial is that two or more people have genuinely joint concerns and values. They recognize a set of common aims, commit to acting as a group to pursue them, and care about the group itself, as an aspect of their own agency. In this way, they create and act from a new, unified entity—a plural agent—alongside their own individual selves.

Helm argues that those who regularly form a plural agent in important and extensive areas of their lives can form a plural *person*. This happens when they deeply identify with one another, and their relationship is central to who they each individually are. It's likely what Aristotle had in mind when he referred to a close friend as "another self."

Closely intimate twins are a compelling example of a plural agent, if anyone is. As one writer on twins notes, "in the best instances" they possess "absolute mutual trust, a highly developed theory of the other's mind, and an ability to work together that surpasses that of any other human dyad." In myths and stories such twins are always pairing up on campaigns and quests: stealing thunder and lightning from the gods in the Pacific Northwest, galloping across the sky on horse- or camelback in Sparta and Syria, transforming into other substances to fight villains in Hanna-Barbera cartoons.

I see Julia's and my suburban child selves in these mythical adventurers, and I recognize something else about the idea of plural agency, too. Helm suggests that his account explains why the loss of a close friend can cause such deep mourning. (Montaigne said after his friend Étienne's death, "I was already so formed and accustomed to being a second self everywhere that only half of me seems to be alive now.") In losing your friend, you've lost the plural person you formed together. If you acted as that plural person in wide and deep domains of your life, it's not purely metaphorical to say that part of your own self has been ripped from your chest. This experience highlights that plural personhood isn't only a matter of action, but also of feeling—including maybe the most intense feelings we humans can experience.

When Julia went through the latest of a series of disasters a few years ago, my therapist asked me if I felt any "survivor's guilt"—the sense that I'd wrongly escaped something Julia had suffered instead. I felt a flash of confusion and found myself snapping back internally, *I can't have that if I* am *her!*

I am her—I felt it beating in my chest. *I am her, I am her, I am her.* I didn't care if it didn't make any sense, if I couldn't defend it in philosopher court. *Fuck that therapist*, my chest said, *and fuck everyone except me and Julia.* You're *other*, I'm not. *I'm her.*

I remember feeling this way when I heard that Julia had broken her leg or arm at school. The teachers would try to get me back in the classroom while our parents took her to the hospital, but I wouldn't have any of it. In my memory, I'm a raging beast straining at the leash, while Julia lies crumpled by the monkey bars. It wouldn't actually have happened that way: I was a quiet child, a little philosopher, I barely ever got publicly het up about things. But there I am in my mind, snarling and frothing, tipped at a thirty-degree angle to the concrete. Despite my weak skeleton, I know they can't hold me back: another second and I will have them. *Let me at her, fuckers. She's not going anywhere without me. She is me. I'm her.*

· ·

If you asked me to justify my late-in-life openness to the idea that twins can be something like a single person, I'd probably draw on a combination of the extended-mind thesis and the joint-agency idea. But something about this troubles me. To ground shared personhood in essentially intellectual acts of thinking, intending, committing, and emoting is to sign up for the idea that personhood is exclusively located in the mind, no? My lifelong preference for that metaphysical position now strikes me as more than a little suspicious.

Is it any surprise that someone with an unpredictable physical disability, and a philosophy PhD, raised in a Western culture, would try to justify the existence of shared personhood by appealing solely to facts about the mind and the will? Maybe we shouldn't think that shared personhood can be bought this cheaply, via some outsourced thinking and joint action. Maybe the body matters more than that here; maybe it's not irrelevant or dispensable to who we really are. I'm much more sympathetic to this objection than I used to be, but it's been painful getting here.

Osteogenesis imperfecta type one tends to take a U-shaped trajectory. Lots of fractures leading up to puberty, a stay of sorts in your twenties and thirties, if you're lucky, and then rapidly increasing breaks as you move into your fifties and beyond. Julia and I are in our midforties, presumably both near the bottom right of our personal U's, but Julia's letter has a shallower dip than mine. Over the past twenty-five years I've had less than five fractures and have usually passed as able-bodied. Julia has clocked up another twenty breaks, and her disability has been more visible, partly because of the OI-related back condition she developed at the age of eighteen. Sometime in her earlier teens, one of her vertebrae secretly slid slowly out of place, like an iceberg creaking off the main shelf. She could have surgery to fix it if she wanted, the doctor told us when delivering his diagnosis, but that would be dangerous. So she held off, and over the following years her spine broke, effectively, into two separate pieces.

For Julia, unlike me, the option of faking a normal body sailed a long time ago, which might explain her comparatively cavalier attitude to her physical self. She recently had ABSOLUTE WRECK printed across her favorite sweatshirt, and she tends to ignore the medical advice of her specialists, a rebellion she calls "exercising agency with respect to my health care." But even before Julia's and my experience of OI began to diverge, she was always much more at ease with her body than I was. It's one of the two or three major differences between us, though not necessarily visible to the eye because I try to pass as chill and at this point in life am a prize faker.

This contrast came up unexpectedly at a writing workshop I attended a few years ago. I'd submitted a memoir chapter about the desperate crush on my philosophy professor that had wracked my senior year of college. The chapter included a lot of material on my lovesick daydreaming and the intellectual history of the Romantic movement, and not a lot on how I wanted to sleep with him or anything.

"I'm not getting a strong sensory picture here," one of my fellow workshoppees mused. "Like, it's all pretty abstract. Can you say more about what this experience felt like sensually? Like where all this landed in your body?"

No, I can't, Christine, I retorted internally. For one thing, I couldn't remember; for another, eww.

"I don't know," someone else said kindly, coming to my rescue. "I feel like that wouldn't be consistent with this narrator. She's very in her head, and it makes sense that the way she writes reflects that."

"Right." I nodded. "This . . . narrator . . . is like that."

"Well, maybe you could get the Julia character to do it for you," someone else suggested. Eighteen-year-old Julia had appeared in the chapter dating one boy, lusting after another, and with a small trail of heartbreak already behind her. "She definitely doesn't have that issue."

"Yes!" several people said. "We want more Julia!"

"Julia is the sensual center of the book," someone remarked.

When I told Julia about this on the phone later, she was very pleased.

That period where I could blithely ignore my body is now well and truly over. A few years ago, my own spine started to revolt and veer toward Julia's in decrepitude. I suddenly developed a set of unpleasant neurological symptoms in the lower half of my body: aching legs and feet, pelvic discomfort, and a fluctuating menu of diverse types of pain throughout my mid and lower back. I was diagnosed with a rare spinal nerve disorder, probably linked to my low-grade connective tissue. I had two rapid rounds of surgery, neither of which helped much, and now I live with chronic pain that flares up and down but never fully goes away.

As a result, I've been dragged, kicking and screaming, toward the acknowledgment that I have a disability, which, in my case, hasn't been too far from acknowledging that I have a body. Once I could sit blissfully typing for many hours, feeling totally abstracted from the physical world. Now my day has to be planned around how long I can sit upright, stand, or walk without becoming distracted, irritable or breathless. My mind and body are in constant tense negotiation over how much time I get to spend doing what I love. How could I believe that the first is at the true helm of me when the second so often wins?

• •

It's easy to think that, once you identify with your body, the prospects of sharing personhood with another human dim. For those of us who aren't conjoined twins, and who don't sign up for the Wari view that personhood tracks exchanges of physical substances, embodied personhood seems to mean metaphysical isolation. But another line of thought suggests that acknowledging your embodiment is a way of increasing rather than decreasing your sense of connection to others.

An important part of my recent spinal trials is that I've become more dependent on and vulnerable to other people than I've allowed myself to feel since I was a kid. My dad flew over from New Zealand to nurse me after my surgeries, and my friends made me dinners, did grocery runs, offered me space on their couches, and accompanied me to medical appointments I dreaded. I routinely have to request rides, ask people to carry my suitcases, allow medical professionals to scan and palpate me, and rely on strangers to accommodate me when I lie down in public places. I used to sail through the world as if my reliance on others were optional and temporary, a lifestyle/leisure choice rather than a necessity. Haha! Nope.

I've since learned that this connection between owning your physicality and owning your dependence on other humans is a common experience for people with disabilities, and an important theme in feminist theory, too. For most of the human past, feminist scholars tell us, across most of the planet, personhood was understood as grounded in social connection. Who you were, in this view, was a function of how you fit into a highly interdependent network of kin and communal relations. As the philosopher Georges Gusdorf describes the picture, "The individual does not oppose himself to all others, he does not feel himself to exist outside of others, and still less against others, but very much *with* others in an interdependent existence that asserts its rhythms everywhere in the community . . . The important unit is thus never the isolated being—or, rather, isolation is impossible in such a scheme of total cohesiveness as this."

But in Europe, under the influence of capitalism, Protestant Christianity, the Enlightenment, and Romanticism, this social picture of personhood was progressively replaced by a much more atomistic vision. From the late Middle Ages on, the best human life has been portrayed in the West as one of self-governed individual action, free of the influence and demands of others. This ideal is metaphysical as much as ethical. Your status as a full person is grounded in autonomous agency, and that requires immunity to interference from whoever isn't you.

If you're a white, able-bodied man at the top of the social hierarchy, it's easy to fool yourself that such radically individualistic autonomy is possible. You're socially set up to ignore the ways your advantages, and your bare existence, depend on the unacknowledged contributions of many others, including women and racial minorities. You're also taught, as a male Westerner, to associate yourself primarily with your mind: it's female, nonwhite, and disabled people who are seductively, aggressively, or disgustingly physical, not you.

Those of us who don't have that kind of privilege find it harder to ignore the ways our lives are socially enmeshed, and, relatedly, the needs of our bodies. If your physical self is functionally impaired, under threat, enslaved, or nursing a baby, it's difficult to stably maintain a picture of yourself as a disembodied reasoner making rational choices in the social void. Your dependence on and vulnerability to others, both mentally and physically, is very much in your face. What those of us who experience this are seeing is a truth not just about ourselves, but about everyone.

I've begun to think that my lifelong resistance to the idea that twins share personhood is connected to this culturally specific ideal of autonomy that I now see as ideologically suspect. When people imply that twins are somehow one person, what they often seem to mean is that twins are *less* than one person: neither of them achieves full personhood by virtue of their overly close enmeshment with each other. ("It's high time you quit being twins and began being people," says one sister's boyfriend in the 1964 teen romance novel *Double Trouble*. "Separate people.") It's natural for us twins raised in Western cultures to react poorly to having our personhood questioned, since our society tells us that we're not real moral agents, rights-bearing citizens, or beings of dignity and worth if we're not full persons. Being half a person, we know, is like being no person at all. But what if this idea that personhood requires closely guarded separateness from others was wrong all along? If so, sharing personhood with your twin could be something to celebrate, not deny or be ashamed of.

Getting to the point of seriously entertaining this idea has been harder for me than Julia. As well as being the sensual center of my memoir, she's the extrovert of our duo, as I've mentioned. On a good spinal day she'd merge her personhood with practically anyone. I prefer to conduct my social life as a series of intermittent and time-limited forays, after which I run back screaming into the forest. (When giving me romantic advice on a bus recently, Julia said, "You don't have to date people like yourself all the time, you know. *You* can be the weird, difficult, and uptight one. I mean, that is definitely possible for you!") For most of my life, my temperament and culture probably combined to make the idea of separate personhood highly appealing for me. Then my body took the mic.

• •

Now I find myself pulled in two opposing directions. Toward the idea that our bodies are essential to our selves, which suggests that my twin and I are fully separate people. And toward the idea that the cognitive, active, and emotional connections between Julia and me—and between many other close duos—are significant enough to make that separation, at best, incomplete.

I still resist the suggestion that Julia and I are simply the *same* person. That would imply that if Julia committed a crime, there'd be no moral difference between punishing her for it and punishing me. And, similarly, no moral difference between my employer paying her my salary instead of giving it to me. Also that I'm her kid's mother, rather than her kid's aunt, and that whoever I'm dating, she's dating, too. Pure chaos!

How can I reconcile my sense that my self is both separate from Julia's and shared with her? Lately I've been thinking that the problem comes from seeing personhood as unitary and static. What if it's dynamic and discontinuous instead? More specifically, what if a person isn't only something you *are*, but also something you *do*? Since what you do varies over time, the boundaries of your personhood could then expand and contract over time, too. In particular, maybe you can move in and out of shared personhood with another person, at different times and in different domains of life, and to different degrees, depending on how you're interacting with them.

In this picture, when Julia is off committing her crime and I'm off earning my salary, our entwinement is minimal, and we don't count as sharing personhood much at all, especially not in a way that's relevant to how we should each be punished or compensated. But when we move into a mode where we're thinking, acting, and feeling jointly, it makes sense to view us as a plural entity, regardless of our physical distinctness. Still, since the scope of operations of our plural entity only extends to particular parts of our lives, its existence doesn't imply that whenever we merge, I'm suddenly, say, the mother of my niece or Julia the partner of my significant other.

If this is the right way to think about it, twins can be both a single person and distinct people without contradiction, and different pairs of twins can be a single person to different degrees. This feels true to my experience, as someone who's been living in a different country from her twin for two decades. Though I still have moments when I vividly feel my personhood merging with Julia's, they're fewer than those I had as a kid, when our everyday lives were much more deeply enmeshed. But when I do have those moments, our having physically distinct bodies, located in different hemispheres, seems to matter hardly at all. I am partly my body, I accept that now. But I'm also part of the unit my young cousin used to call Jua-Lena—a being whose boundaries stretch well beyond my skin, sometimes to the farthest reaches of the planet.

• •

This business of cleanly demarcating where one person starts and the other ends: What's it all about? These days I can't resist the idea that to try to securely wall off your mind from your body, and securely wall off both your mind and your body from other people, is to try to protect yourself from what you can't control, from what might hurt you. Twins and disabled people—and, especially, disabled twins—fascinate and disturb us because they vividly model the breaching of these walls that many of us—consciously or not—strive to erect. It's a valuable lesson, because if the social conception of personhood is onto something, this whole project of sharp interpersonal demarcation is conceptually bound to fail. If our social connections are part of our identity, if to be a person is to be in relationship with other humans, as the playwright Tony Kushner puts it, "the smallest indivisible human unit is two people, not one; one is a fiction."

Still, I don't want to make all this sound too easy. Some breaches of personal boundaries really are harmful, and it's far from clear that whatever's gained is worth the cost. The risk is greatest in relationships with significant inequality in power. The idea of interpersonal union may sound appealing in a well-balanced, loving twinship such as Julia's and mine, or a friendship between Montaigne and his equally aristocratic male friend Étienne ("Neither of us reserved anything for himself, nor was anything either his or mine"). But what often happens when non-equals merge is that one party gets a much better deal.

Romantic relationships between men and women in a sexist society are the obvious example. As Susan B. Anthony put it, the law in her time effectively held that "husband and wife are one, and that one the husband." The result, among other things, was that wives couldn't keep their own wages, enter into legal contracts, or sue their husbands for rape. Things improved legally in the twentieth century in much of the world, but it remains the case, informally, that women on average do the lion's share of the housework, childcare, and emotional labor in a straight partnership, that their career plans and personal projects take a back seat to their partner's, and that they're at much greater risk than men of partner violence and abuse. Gloria Steinem once described marriage as "an arrangement for one and a half people"—an improvement on the math Anthony was pointing to, but not much of one.

The risk of domination in relationships involving a fusion of selves extends beyond the gender case. One twin I talked with about this question said she felt smothered throughout her childhood by her overbearing twin and could only breathe freely when they put an international border between them. An Indian friend who was with us also gently suggested that people who wax lyrical about non-Western conceptions of personhood often don't have much personal experience of living in a non-Western culture. I'll take it!

In explaining what's troubling about the oppressive examples of interpersonal merger, the modern commitments to individual autonomy and interpersonal boundaries that I've been criticizing turn out to be helpful. A feminism that views the liberal attachment to the separateness of persons as fully misguided is apt to shoot itself in the foot.

If I'm focusing on the benefits rather than dangers of relationality, it's because I think many of us raised in Western societies, myself included, continue to overemphasize the latter. Reflecting on the case of twins can help us individualists adjust the balance here and give us hope that we might get the damn thing right in our own lives. Alice Dreger notes that conjoined twins who survive infancy are faced more starkly than any of the rest of us with the challenge of negotiating mutuality and independence. Yet somehow they manage to do it. By necessity, they learn to cooperate closely in just about every domain of life, but they maintain their distinctive identities throughout. Moreover, Dreger reports, they tend to see their conjoinment not as a threat to their individuality but as an integral aspect of it. Being conjoined is part of who they each are and hardly ever something they want to eliminate. (While most of nineteenth-century America assumed that Chang and Eng Bunker wanted to be physically separated, the twins themselves used to cry when it was even mentioned.) Dreger reports that it's surgeons, not conjoined twins, who urge that even highly dangerous separations are necessary, to secure the twins' autonomy. The twins themselves believe they already have it. What makes the rest of us so sure they don't?

• •

At the age of thirty-four, to fix her escalating pain and prevent serious neurological dysfunction, Julia had surgery to fuse the two parts of her spine with a titanium plate. The operation took eight hours, twice as many as it was meant to, because the surgeon struggled to find any bone inside her that was solid enough to fix the screws into. When one screw broke right through a vertebra and came out the other side, he almost gave up and considered coming in from the front, which would have required puncturing a lung. I was in my apartment in Massachusetts while this man sliced up my twin in Belgium. I lay on my couch for hours, in the fetal position, waiting for news. I felt like I might be dying.

Julia survived and so did I, but I sometimes wonder what will happen when, in the future, one of us doesn't. The sensible answer is that the one who's left will experience acute and then chronic grief that quietens over time. The less sensible answer is that the second of us, too, will be snuffed out instantly, as a simple matter of logic. For all my philosophizing, some prerational part of me doesn't rule that latter option out. I can't imagine life without Julia, not just in the sense that I couldn't bear to experience it, but in the sense that I literally *can't imagine* it. She's been there from the very beginning, since our primordial breakfast, when that single egg cracked in two. You can tell me that I'm fundamentally separate from her, that we're not metaphysically conjoined, that my personal boundaries are and should remain intact. You can tell yourself that, too, about you and everyone you love. As for me, I'll believe my independent existence when, God forbid, I see it.

ARE YOU TWO IN LOVE?

The first time I slept with my future husband was in a beach house an hour north of my childhood home. My boyfriend had been given the keys to it for the weekend by its owner, Camilla, a retiree he'd worked for in the New Zealand government. The house was a modest rectangular modernist box, raised off the lawn, wood paneled inside, except for floor-to-ceiling windows at the front with a wide view of the coastal sky. Neither the building nor its contents seemed to have been updated since the sixties, and the atmosphere inside was super cozy. A pair of recliners upholstered in nubby beige angled toward each other in the lounge, around a low table crying out for matching teacups and scones fresh from the oven. When my boyfriend mentioned that Camilla's sister was often at the house with her, I instantly saw them sitting there, two old maids in cardigans and sensible shoes, tut-tutting over the papers and arguing over who'd pick up the cold cuts for their afternoon guests.

My boyfriend and I had only just started dating, but he was handsome, smart, and kind, and I was optimistic about our future together. I was also, technically, still a virgin and knew I was likely to rectify that within the next half hour. As I moved through Camilla's house, I should have been consumed with excitement, capable of attending to nothing but our two young bodies and what we were about to do with them. Instead, a different thought was forming in my mind. It reached its climax as I laid my valise in the sunny bedroom, where my childhood would soon officially end. *Omg*—my chest thrummed—*what if Julia and I bought this place when we retired and lived here together till we died?*

I earned my adult badge, had a nice weekend, and told Julia about my vision as soon as I got back down the coast. She signed up for it instantly, as I knew she would. Whenever bad shit happened in the years to follow, one of us was likely to sigh, "Won't it be so relaxing when we're living at Camilla's?" Car accident, dislocated foot, rejected manuscript, extramarital infatuation—to cheer us both up, I'd remind Julia of the comfy chairs overlooking the sloping lawn, how the late-afternoon sun sent a gentle glow round the room, what a perfect place it would be for knitting and reading, while our baking rose contentedly in the oven and our aged ankles swelled companionably in our slippers. After Julia got pregnant, we imagined her fetus arriving to deliver us groceries, introduce us to its partner, and clip the backyard roses with its fully developed hands. Our husbands were trickier to manage: it was hard to envision them on-site. But they were a decade older than us, and male life expectancy is lower. We were hoping, I guess, that they'd be dead.

• •

Later, in the midst of my divorce, I began to see all of this as not having been a super-great omen for my marriage. When dragging boxes from the conjugal home to my car, for transportation to my sad little rental across town, or when exchanging crippling glances with my soon-to-be ex-husband in the offices of therapists or lawyers, I'd sometimes think of my blissful vision of twin retirement at Camilla's with a more critical eye. "Jesus, Lena," I recall puffing to myself once midstairs, while re-homing my half of the liquor cabinet, "what was *that* about?"

One thing Camilla's offered to my younger self was a reassuring existential guarantee. Even if this new kind of intimacy with my boyfriend didn't work out, I was likely telling myself, with more prescience than I knew, I'd always have a backup in place: a permanent home by Julia's side. The idea that twinship provides a uniquely secure solution to loneliness and alienation is probably the biggest attraction of being a twin, for twins and singletons alike. One woman in a collection of photos of twins, Judy, reports, "Even if you're unsure of God, you do know that, because of your twin, you are never alone."

On first reading this, I wanted to say, "Judy, that's so schmaltzy!" But I have to admit I've experienced a similar feeling, and I'm not sure what my emotional life would be without it. (How do singletons live?) I can count on Julia to want to hang out with me, but the reassurance she provides goes beyond that. Some part of me seems to believe that the mere fact of Julia validates my existence. She makes me feel my membership status in the universe is active, as if I've already passed some crucial cosmic test and every later qualification is optional. Just thinking of her calms me down, the way I imagine the thought of God, Gaia, or eternal flux does for believers, mystics, and Buddhists.

Alongside that, Camilla's represented, maybe, a return. Julia and I began our lives together in the snug, dimly lit, twentieth-century lounge of our mother's womb. Wouldn't it be perfect, I imagine my younger self thinking, wouldn't it just be *right*, for us to end up that way, once all the stressful subplots of life had played themselves out?

Though I didn't know it that weekend at the beach, this conception of the ideal relationship as a reunion of two parts once sundered has a long history in Western culture. It appears most famously in the creation myth told by Socrates's fellow dinner guest Aristophanes in Plato's *Symposium*. We humans were originally egg-shaped, Aristophanes says, with two faces, four legs, and four arms, which allowed us to tumble head over heels at speed. It was great for a while, but made us too powerful and arrogant for the gods' tastes, so Zeus split us neatly in two. Ever since, we've felt existentially bereft and have yearned unhappily for reunion with our other half. We can never fully manage it, but we can come pretty close if we're lucky enough to relocate our primitive soulmate, somewhere out there on the planet, equally lonely and yearning for our embrace. "And when a person meets the half that is his very own . . . then something wonderful happens," writes Plato, "the two are struck from their senses by love, by a sense of belonging to one another, and by desire, and they don't want to be separated from one another, not even for a moment."

The idea of ideal love as a terrestrial twin was driven underground by Christianity—human hearts were meant to find their rest in God—but resurfaced and reached its peak in nineteenth-century Romanticism. Shelley described a couple in love as "one soul of interwoven flame," and in *Wuthering Heights* Cathy exclaims to her nanny, "Surely you and everybody have a notion that there is or should be an existence of yours beyond you . . . Nelly, I *am* Heathcliff!" The core features of our contemporary ideal of romantic love were born in this period: the ideas that love between romantic partners, at its truest and best, is unconditional, selfless, and eternal, a haven of peace, a source of ultimate meaning, a guarantee of perfect happiness, and a redemption of all life's sufferings. If romance subs for God in each of these ways, it also promises us the closest thing to immortality we skeptics can now hope for, by extending each lover beyond the bounds of their own mortal frame. "What were the use of my creation," Cathy asks, "if I were entirely contained here?"

Seen against this backdrop, my vision of a beatific twin reunion at Camilla's seems distinctively Romantic, while at the same being very unromantic, as far as my future ex-husband was concerned. This reflects a broader problem with the relationship between twinship and romance, which is an awkward fit by any measure. Twins clearly function as metaphors for singleton romantic relationships in the Platonic myth, and as Alice Dreger points out, many a love song can be read as a covert analogy to conjoined twins. ("I've got you under my skin," Sinatra croons to his lover.) But even Plato makes his myth's uneasy fit with singleton romance quite clear. Our reunited ancestors "cannot say what it is they want from one another," his mouthpiece, Aristophanes, states. "No one would think it is the intimacy of sex—that mere sex is the reason each lover takes so great and deep a joy in being with the other. It's obvious that the soul of every lover longs for something else."

Aristophanes suggests romantic lovers yearn to fit together not merely sexually, but in a more thoroughgoing and permanent way, like two lifelong halves of an apple. They want what's impossible for them, we might say: to be actual twins. Identical twins, at least, literally come into the world as the result of asexual reproductive division, just like Aristophanes's split ovoid creatures, or Eve, created from Adam's rib in the Garden of Eden. Spend more than a little time thinking about that, and it starts to seem that we identicals don't so much animate the Romantic ideal for singletons, as underscore the impossibility of their ever truly attaining it.

• •

In Western culture, the myth of twinship—*actual* twinship—as the ideal human relationship takes its most famous form in the story of Castor and Pollux. The two were said to have resulted from their mother sleeping with Tyndareus, the king of Sparta, and Zeus, the ruler of Olympus, one after the other, resulting in one ordinary human baby and one touched by divinity, delivered at the same time. As the story goes, Castor and Pollux grew up to be the best of friends and spent all their time with each other, wearing their matching pointed felt caps, said to represent the vestiges of the two eggs from which they'd hatched. They were excellent horsemen and loved hunting, sailing, boxing, and merrily picking fights with their rivals. Eventually, though, the fun ran out when Castor was mortally wounded in a brawl. As Castor lay dying, Zeus informed semidivine Pollux that he could give half his immortality to his brother, which would mean sharing half of Castor's death. The details of the offer are fuzzy: in some tellings, the twins would live on alternate days forever; in others they'd winter together in Hades and summer on Olympus.

According to the ancient Greek poet Pindar, Pollux "didn't give it a second thought." In a choice between immortality and Castor, he went with Castor—of course. He likely felt about it the way Cathy felt about Heathcliff when she said, "If all else perished, and *he* remained, *I* should still continue to be; and if all else remained, and he were annihilated, the universe would turn to a mighty stranger: I should not seem a part of it." Deathless though he was, Pollux couldn't live without his twin.

When I talk about my relationship with Julia, I find myself speaking in phrases sapped of force by centuries of overuse by singleton romantics. *What does it feel like?* people want to know. *It feels like she's always there for me,* I say. *I can never get enough of her company. We have more fun together than we have with anyone else. We understand each other better than everyone. We can tell each other everything. We trust each other completely. We'd do anything for each other.* I make myself sick. But now and then the violent truth underlying these clichés surfaces, giant and graceless, like a whale.

I was speaking about Julia a few years ago to a minor acquaintance at a bar when I heard myself say casually, one sip into a martini, "If I betrayed her, I couldn't live." I blinked and returned to my drink, self-startled. It wasn't a thought I'd explicitly entertained before, let alone articulated to a semi-stranger. Like Pollux, I recognized it instantly as the basic fact of my life.

. .

The cultural heritage of the West contains, then, two visions of the perfect human relationship: one centering romantic couples, one—lower profile, but still potent—centering twins. Signing up for both visions at once is awkward because each has exclusivity built in. We don't think of triplets, quads, quins, or sextuplets as enjoying the ideal human bond because their relationships raise the possibility of divided loyalties, conflict, and secession. Likewise, our mainstream vision of romance doesn't countenance polyamory, or even serious serial monogamy. There's only one person out there in the universe to twin your heart to, the traditional story goes, though ideally it'll take a series of thrilling dalliances with hot pretenders for you to find them.

Society's standard solution for this conflict is chronological. People love to see young twins acting like mini Castors and Polluxes. When Julia and I were released a few feet from each other on our first walk in a public park, we brought a neighboring man to tears by instantly tottering back toward each other, our four arms outstretched. But we're encouraged to see young twinships much as we're encouraged to see intense child and teen friendships: as practice for the real thing, which is a grown-up, romantic relationship between unrelated adults. When the time comes, the idea goes, twins will and should shove aside their sweet little bond, so that the serious business of marriage and parenthood can take center stage.

That's a nice, tidy fix, and no doubt many twins pull it off. Good for them. My own experience—one shared, I take it, by many other twins—is off script. My husband was never serious competition for Julia, and none of her romantic partners were ever serious competition for me.

Who is this guy? I remember wondering, while attempting small talk with Julia's first boyfriend, a tall, baby-faced Italian she met at seventeen. *Why is he here?* I didn't know what to do with him, what his function was, where he was supposed to go. Julia reported confidentially that her beau harbored some kind of amorphous hostility toward me, and in the absence of her diverting presence, I occasionally saw a spark of it flash out of his otherwise syrupy brown eyes. What, was he *jealous* of me? I marveled. "LOL," I would have exclaimed, had it not been the pre-texting nineties. He reminded me of that mouse in *The Chronicles of Narnia* trying to attack a giant with its toothpick-size cutlass.

We got rid of that one in due course and proceeded as before. Things became more complicated with Julia's second boyfriend because I'd inched forward in my attitudes to young men by then and had a crush on him myself. Julia and I narrowed in on him simultaneously, when introduced to him by mutual friends. It was clear one of us was going to get him, but not clear, for a few weeks, which. I vividly recall the encroaching sense of doom of that hot and heady month. Julia and I would position ourselves opposite our prey in friends' houses, in bars and cafés, on lawns and at picnic tables, and flirt with him in tandem. I tried my hardest, but Julia's talents were vastly superior. She'd had more experience, and her personality allowed for a brashness and social persistence that mine couldn't match. We were like twin Love-a-Lot Bears busting Care Bear Stares from our respective bodies, but hers was a giant resplendent rainbow arcing directly into our crush's heart, while mine fizzled out unnoticed half an inch from its point of exit. As the days progressed, I felt the energy coalesce around the two of them and resigned myself to my fate, which included loaning my leather jacket to Julia for their first date.

It soon became clear that the right choice had been made, and my own crush vanished. The three of us spent a happy two years going on road trips together, attending concerts and parties, cooking experimental dinners, and working complex in-jokes to death. I felt as if I'd gained, if not a triplet, at least a brother. Eventually Julia came to see the situation that way, too, and the romantic phase of the relationship wrapped up.

Julia's husband was next-level, but even he caused me little concern, maybe because he reminded me of mine. Both husbands had graduate philosophy degrees, worked as corporate consultants, and were enthused by wine, science, and Bach. They were older, taller, and more professional than Julia and me: we were grad students when we met them, a few years apart, and we looked and acted as if we were barely out of our teens. The husbands took us on fancy trips, prepared elaborate meals for us, and were extremely quick and well read. They seemed different from us in many ways, but they matched each other in many others, the way Julia and I did. Marrying them didn't alter Julia's and my relationship at all. In some ways it was like engaging in interesting joint parallel projects, the kind that allowed us to usefully compare notes. We didn't particularly bond with each other's husband, and our bonds with our own turned out to be fragile. Shortly after our twin divorces, which happened a few years apart, we decided to refer to ourselves as Calamity and Catastrophe—and then we laughed, as if in some obscure way we'd won.

The thought of our victory was funny because we knew *almost no one would agree with us*, and the joke gained its edge from our having entered dangerous territory. We were like romantic outlaws—us against the world!—except that it was romance itself whose bounds we'd stepped outside of. Twins who make that illicit move tend to find the admiration and envy singletons lavish on them twist into something new.

• •

A year or so after her divorce Julia came over to Boston for a summer vacation. On the second day of the trip, we were sitting on my couch, with our twin cups of tea lined up on the coffee table, settling in, I thought, for a relaxing chat. But when I looked over at Julia, she seemed on edge. I was surprised: she seemed to have been happy recently, enjoying time with her toddler, embarking on new research projects, and expanding her social scene. I'd just asked, in fact, about a new friend she'd made through her daughter's nursery school. The friend seemed cooler than the standard crop of people Julia had been running into since moving to Luxembourg a few years ago: rich couples with conservative leanings and oppressive furniture. This new one was part Italian, with cropped dark hair; she'd worked in the film industry and zipped around town on a Vespa. She was a strapped-for-cash divorcée with a teenage lesbian daughter, frenetic hand gestures, and a crazy laugh.

"I really like her," Julia was saying.

"That's great!" I replied. "Finally you can have a proper friend over there!"

"Yeah, I guess so," Julia said.

"What?" I asked.

"I feel stressed about it," she said. "Like I'm not sure how she feels about me and I just . . . I feel stressed about it."

I shifted gears, backed up, and asked some tentative questions, but got nowhere. Julia couldn't seem to articulate even to herself what was distressing her, so we moved on to other things. I forgot about it, till Julia phoned me in the middle of the day a month after her return to Luxembourg. She'd realized, she announced, she was in love with the Vespa rider.

"Oh!" I exclaimed. "Okay! Wow!"

I was very surprised and also trying very hard not to sound surprised, since this was—what? This was a coming out situation? Julia was coming out to me? *Oh my God! What?! Julia was in love with a freaking woman!*

At that time my three closest friends were lesbian or bisexual, along with maybe half of the rest of the people I hung out with. I worked at a women's college that's a queer magnet, and I split my time between Cambridge, Massachusetts, and San Francisco, California, two of the most LGBTQ+-friendly parts of the country. So many queer or queer-adjacent people were around, I'd started to feel like the straight one out. Given all this, I was pretty sure I didn't have an ounce of homophobia in my body. But implicit prejudice is a thing, and I didn't want any nasty drop of it to come out inadvertently over the line. Thankfully I was running late for a haircut, so I was able to cut off the call pretty swiftly and try to recalibrate solo.

"*My twin just came out to me!*" I announced to my stylist, Derek, as he attached the cape around my neck with a flourish.

Derek was not as interested in this news as I'd hoped. Maybe he was bored of coming-out stories or worried, like me, that I was a closet gayphobe. I settled back into the chair and tried to avoid eye contact with both Derek and myself. But the mirror beckoned.

Julia is gay! my eyes blasted.

What can I say, it felt like kind of a big deal.

Julia has tended to embark on both minor and major life changes in advance of me. She worked out how to blow bubble gum first, she beat me in putting her head underwater in the swimming pool, she got her driver's license a year ahead of me, and she reached all four bases before I'd even arrived on the sexual playing field.

"Maybe you're next!" she joked over the phone a week or so after her revelation.

"Haha!" I said.

It seemed exceedingly unlikely, and I spent barely a second thinking about it. Two summers later, I fell in love with a woman myself.

. .

Alongside, and in uneasy tension with, the romantic vision of twinship lies a starkly opposed pathological one. In this alternative view, twin bonds are highly dysfunctional: sick, morally corrupt, self-destructive, and doomed. An interesting feature of the pathological vision of twinship is that it lines up in several ways with the picture that straight culture often paints of queer romantic relationships. Maybe that's why some fictional twins—even the allegedly straight and different-gender ones—come across as super gay.

Edgar Allan Poe's "The Fall of the House of Usher" stars Roderick and Madeline, a pair of genteel twins who live together in their crumbling ancestral manor beside a remote mountain lake. Our narrator comes upon Roderick there, pale, gloomy, and distraught about his sister, who suffers from a mysterious illness that causes her to wander the house in a trance. Madeline is soon found dead, and the narrator helps her grief-stricken brother inter her, cheeks still warm, in the family vault. In the days that follow, the creepy atmosphere ramps up: strange noises, alarming weather, and an odd glow arising from the lake. Ultimately, Roderick reaches hysteria, sounds of clanging and shrieking are heard from deep in the house, and Madeline reappears at the bedroom door. She and her twin fall to the ground in a deathly embrace, and the singleton narrator—super over it at this point—takes flight. Turning back as he runs, he sees the moonlit manor cleave in two and disappear into the lake.

Throughout this ordeal, Roderick pulls off the impressive feat of coming across as both hot for his sister and stereotypically gay. He's aristocratically idle, tremblingly sensitive, and artistically inclined; when he can work up the energy, he makes fan art of haunted houses. Physically, he has lips "of a surpassingly beautiful curve," a delicate nose, hair of great softness, and a tremulous voice, and he closely resembles his female counterpart.

Thomas Mann's short story "The Blood of the Walsungs" features another pair of male and female twins,

Siegmund and Sieglinde, named after the incestuous characters in Wagner's *Die Walküre*. The two attend a family dinner at which they persistently seek out each other's gaze: "Their eyes met, melted, formed a rapport to which no one else had access or entry." Then they head out for a performance of Wagner's opera, which throws them into a sensual frenzy that ends with them ravishing each other on a bearskin rug back at the house. Siegmund Walsung is coded as spectacularly gay. He has full, soft lips and long slim hands "no more masculine than" his sister's. He takes an hour to dress for the opera, wears pink underwear, and is very into flowers. Though he dabbles in painting, any serious artistic vocation is ruled out by his lack of talent and listless, superficial personality. Both Walsung twins appear unwell: they're alternately lethargic and feverish, and Mann compares them to "self-absorbed invalids." Neither of them dies at the end of the story, but the whole piece is suffused with a sickly, overheated atmosphere that suggests the inevitable outcome can't be too far off.

What is going *on* here? For one thing, an association of both twinhood and gayness with immaturity. The bodies of the Walsung twins are "childlike despite their nineteen years," and they've clung to each other continuously "since they had both babbled their first sounds, taken their first steps." Siegmund has zero interest in an actual career, and Sieglinde is in no rush to move on to wifehood, despite having a fiancé already lined up across the dinner table. The Usher twins simply refuse to leave their childhood home like ordinary adults and exercise their responsibilities in the outside world. As a contemporary psychologist might diagnose the problem, both sets of twins have clearly failed to individuate. They've used each other as mother substitutes, "transitional objects" from which they've failed to transition. Deprived of the fresh air of outside company, their closed dyad has turned in on itself, intensified, stalled, and decayed.

Queer couples have often faced this same charge of arrested development. They've been assumed to be stuck in perpetual adolescence, by virtue of their pathological narcissism and lazy refusal to sustain straight social structures, including, until recently, the reproduction of the next generation. Freud viewed same-sex desire as a failure to mature beyond infantile oral and anal fixations. Such forms of intercourse, he wrote, "are ethically objectionable, for they degrade the relationships of love between two human beings from a serious matter to a convenient game, attended by no risk and no spiritual participation." Quentin Crisp, gay icon, said of the 1930s Britain of his youth, "an effeminate homosexual was simply someone who liked sex but could not face the burdens, responsibilities, and decisions that might crush him if he married a woman." Gay men in particular have tended to be portrayed as irresponsible, decadent partiers, motivated mainly by hookups, cocktails, and drugs, but lesbians, too, are treated as socially immature, stalled at the sisterly stage of female relating.

Another queer-adjacent trope we see in the Poe and Mann stories is an association of twins with the unnatural or inhuman. Twins and doubles are popular in Gothic fiction and horror films because of a widespread sense that they disrupt the natural, i.e., singleton, order. They're often compared to nonhuman animals, maybe because they're born in litters. Siegmund and Sieglinde "pla[y] like little puppies" and return from the opera "bundled up so warmly and delicately, the two strange, dark creatures." When Sieglinde turns to her fiancé for the first time at lunch, she peers into his face with eyes "that spoke as mindlessly as the gaze of an animal," clasping "the slender hand of her twin" all the while. The related association of twins and queer couples with sexual perversion is also at play in both stories, though in Poe with a lighter touch. What gets cutesily called twincest has long been a popular theme in porn, novels, and film (think of Rahel and Estha in *The God of Small Things*, Jaime and Cersei Lannister in *Game of Thrones*, and Hanna-Barbera's Wonder Twins, if one notorious comic book cover is anything to go by). In actual life, to do a reality check, twincest is rare. Some twins who were raised apart have become romantically involved without knowing they were siblings, but twins raised together generally report no more sexual attraction to or activity with each other than the average sibling pair does, i.e., none.

How does this pathological plotline tend to play out? Generally not super well for either the twins or the gays. Until recently, stories about queer couples have almost never ended happily. For their part, fictional twins are at risk of permanently losing themselves in a morbid spiral of mutual obsession, Poe/Mann–style (as in David Cronenberg's *Dead Ringers* and Peter Greenaway's *A Zed & Two Noughts*). Alternatively, at least one twin comes to hate the other, due to envy, jealousy, or rivalry. Remus was said to have been killed in a spat with Romulus over the site of Rome; in Genesis, Jacob and Esau fight over their inheritance and the future of the Jewish people. To shift to pulp fiction, in Lisa Scottoline's 2010 *Think Twice*, one sister, out for the other's job, boyfriend, and life, buries her twin alive. (Bennie crawls out, only to find that in her campaign of revenge "she is perhaps more like Alice than she ever realized." Saw that one coming.)

The worst-case scenario is that one or both twins die, either violently at each other's hands or through neurotic dissolution brought on by their suffocating bond. The best-case scenario is that the siblings end up estranged. Esau and Jacob escape their political feud only by founding separate nations, getting them permanently out of each other's hair. Even fictional twins with comparatively peaceful relationships don't get to live happily with each other for long. Though twins in "comedies of confusion" may end up alive, the literary theorist Juliana de Nooy notes they're "usually separated at birth and in plays never appear on stage together till the last scene. Twins who aren't separated early, whose lives are entwined, are frequently doomed." Even Castor and Pollux, the best of twin buddies, survive only partially, enjoying the dubious prize of half a year in hell for eternity.

• •

My own first experience of queer romantic attraction didn't go very well. Like Julia, I'd fallen in love with a straight woman who was a close friend, so things were super awkward and hopeless from the beginning. My gay best friend, Coco, assured me that this was often how it went. You begin with a straight femme, she said, as a kind of starter drug, and next thing you know you're living in an ecofeminist separatist community with a butch with a buzz cut and five cats. I wanted to believe her, but in the meantime I felt terrible. All loved up, with nowhere for that love to go—and, also, maybe I was a little homophobic after all.

I was very into this particular woman, but was I into being gay? Unclear. When I imagined Julia's and my hopefully more successful romantic futures with women we hadn't yet met, I had complicated feelings about it. The idea that neither of us would have to deal directly with the patriarchy under our own roofs was super appealing. Our husbands had been feminists, but they were still cis men, and Julia and I had taken up roles with them that struck me in retrospect as pretty traditional. Imagine ditching that fraught setup for good, I marveled, and having romance be more like—well, like Julia's and my relationship: fundamentally equal and marked by a depth of understanding that maybe just wasn't attainable with a guy.

But, if I was honest about it, that vision of us parallel nesting in our woman-on-woman households also set off a buzz of alarm. What was this midlife tandem conversion really *about*? I found myself wondering. Were Julia and I retreating to intimacy with another female as some kind of cowardly regression, after simply failing at adult relationships with men? If we found these future women, would it feel *too* comfy? Would all four of us get lesbian bed death, take up knitting, Patagonia fleeces, and homemade vegan meats and watch our lives grind to a general halt? Were Julia and I trying to retire to Camilla's three decades early, as a result of just, you know, giving up? Or—worse—were we trying to date *each other* in some fucked-up sublimated way?

With a little distance, I can see these feelings track the Poe/Mann model: more than a little internalized twinphobia to match the internalized homophobia. Twinsies.

The low-key tendency to treat twins as kinda, sorta gay—in the Poe and Mann stories, in the common sexual fantasy of identical twins making out, in my own inchoate

feelings about the matter as a baby gay—is matched by a tendency running in the opposite direction. Writing of queer couples, the literary scholar Joseph Bristow says, "Whatever our differences . . . we have historically been regarded . . . as twins." In the latter case, the association probably springs from overemphasizing the role of gender in defining a person's identity. If the most salient feature of Adrian is that he's a man, and the same goes for Alfonso, the Adrian-Alfonso romance is going to look like a pairing of closely similar partners, the way a twinship does to the average singleton. To state what should be obvious, that's just a mistake.

A different error likely underlies the parallel idea that twins are effectively gay (at least in the sense of the same-gender ones wanting to sleep with each other). Many of us seem tempted by the thought that any highly intense and intimate connection will necessarily tend toward the sexual, or at least the romantic. If you take that view, many twins are going to seem hot for or in love with each other, whatever their expressed views on the matter.

As a twin I find this irritating, but I have to admit I'm not immune to the general tendency myself. I thought about this a lot while I was trying to get a grip on the nature of my relationship with the straight woman I fell in love with. She didn't want to sleep with me, that was tragically clear. But it was also clear that the charged vibe between us wasn't all coming from my direction. She reacted to me more intensely than any friend I'd had before her and said things to me that, coming from any other mouth, I'd have interpreted as romantic. For a while, I tried to convince myself that she had the exact same feelings I did and was just conceiving of them differently. But that didn't seem right: my levels of exhilaration and obsession were definitely higher, and the way she talked about men she'd been in love with was very different from the way she talked about me.

I now think that much of what made this relationship so confusing and insane making for me was my desperate attempt to fit both of our feelings about it into the very limited set of social categories I'd been operating with up until then. Looking back, it seems I'd effectively divided intimate relationships outside the family into only two camps: romance (intense, passionate, and exciting) and friendship (milder, less consuming, and less thrilling).

What a parched scheme! Some romances are low-key in emotional tone, some friendships are ardent, and other close and valued bonds don't fit either of those molds. Some talk of a category, "romantic friendship," that bridges the two—maybe this is what my friend felt for me?—and then there are, say, mentors and mentees, beloved neighbors, creative collaborators, long-term community members—and twins—where the people involved resemble neither romantic partners nor standard friends, while mattering deeply to each other. You could double down here and claim that all intense relationships, twinships included, are unacknowledged sexual relationships. Or you could push back on the blinkered idea that every instance of charged energy has to be erotic.

• •

I find myself totally convinced by all of that, but you know what? I actually do think twins are kind of gay. Or at least, being a twin feels to me like being gay, and being gay feels to me like being a twin, in one major way.

Since coming out, I've experienced a physical sense of relief whenever I run into someone new who isn't straight. I feel my chest relax and an internal vista open up, as if the world has expanded several inches outward. When I spend a lot of time with straight people who aren't my friends, I feel the opposite: a sense of tension and restriction, of being on my guard. Everyone's small-talking about their spouses, their kids, their two-income home renovations, their large family vacations; I could talk about my own single life dating non-men and they'd all respond in a friendly fashion, but it feels like a lot of work. I've spent the huge majority of my life identifying as straight, and as a femme I can pass that way whenever I want, so I'm not trying to host a pity party here. But the feelings of relief and defensiveness are there—and they remind me of something.

I was at a department lunch this past fall, chatting to students about writing this book, when one of them eagerly announced, eyes lit up, "*I* have an identical twin!" And there was that same shift in my chest, the one I feel among fellow queers, and the one I now realize I've felt many times before when running into a fellow twin. I'd only just met this woman, but it was as if we'd carved out an instant miniature community, formed of secret things we knew about each other that were wide and deep. As the singletons peppered us with their predictable queries, the twin and I became impatient. *So over this!* our faces flashed to each other. We were dying to talk about our beloved sisters together, away from those alien and alienating eyes.

This is probably how Julia and I felt at the twins-club events we attended when we were little. We only went to a few because our mother wasn't a joiner and maybe the whole setup felt too cutesy. But I remember having a blast at one holiday party in the summer of 1985. There were child twins everywhere, many in matching outfits, and a pair of men in Santa costumes handing out candy. There were even twin fire engines to climb into, with twin sirens to blast, as if a mad scientist had let rip in the area with a duplication gun. We all ran around screaming, on a combo sugar/twin high. It was as if the animals had broken out of their stalls, two by two, and taken over the ark. For an hour or so, our weirdness made us normal.

The weirdness of twins tracks the weirdness of gay couples in one important respect. The normative expectation in our culture is that highly intimate adult couples will be romantic partners of different genders. Many queer people violate that injunction. But so do many twins, in at least one of two ways. If they're close, they're in a super-intimate nonromantic relationship. And they often share a gender.

It's not obvious why the weirdness of twins should result in so many of us being killed off in myth, literature, and film, but the parallel case of queer couples offers a hint. One common explanation for homophobia is that we're all a little bit gay or at least worried we might be. Maybe in the domain of love we're all worried we're a little bit twinny, too.

Twins aren't the only ones who come across as self-involved when in love. If Dromio of Ephesus says to his twin in *A Comedy of Errors*, "Methinks you are my glass," Justin Timberlake sings the same to his love interest in "Mirrors": "You reflect me, I love that about you." Cathy says she loves Heathcliff "not because he's handsome, Nelly, but because he's more myself than I am. Whatever our souls are made of, his and mine are the same."

Nor is twinship the only bond in danger of being less than impartial, from a spiritual and political point of view. As C. S. Lewis said of friendship, love is "a sort of secession, even a rebellion . . . a pocket of potential resistance." A strong twinship gives a particularly marked impression of self-sufficiency and a private zone from which others are excluded. What better hook for singleton lovers looking to project onto others their concerns about being socially irresponsible?

Finally, any intensely intimate bond will tend to fire up broader human anxieties about engulfment and loss of self in human relationships. It's tempting for singletons to use twins as a teaching tool about the dangers of mutuality and to pin the disastrous instances on them. *i love u babe but i wouldn't like sink my ancestral manor into the lake for u or anything!!!*

• •

Queerness isn't just a matter of being different, but of being different in a way that makes things socially difficult for you. As bell hooks defined it, it's "about the self that is at odds with everything around it and has to invent and create and find a place to speak and to thrive and to live." Twins, as twins, don't face that degree of social pressure, and people in queer relationships usually do. Queerness presents a much more radical challenge to the social order than twinhood does.

But in one specific way twinship, like queerness, represents a genuinely radical form of love. What twinships most centrally depart from isn't *hetero*normativity, but what the philosopher Elizabeth Brake calls *amato*normativity. This is the widespread assumption that centering your life on a single, enduring, monogamous, sexual-romantic relationship is both normal for humans and the uniquely best way to achieve a meaningful, fulfilling, and ethical life. The amatonormative ideal includes queer romantic couples as much as straight ones, at least where queer partners can marry or establish a marriage-like relationship. But it excludes many romantic anomalies, including asexuals and aromantics, people in poly relationships, divorcé(e)s, and persistent singles. And it leaves out all those who center their lives on nonromantic relationships, including the chosen families, platonic partnerships, and extended-care networks that are common, variously, in queer, BIPOC, and senior communities.

Though Brake doesn't mention it, the amatonormative ideal also leaves out many twins. In Arundhati Roy's *The God of Small Things*, the twin heroine Rahel asks her cousin, "Who d'you love Most in the World?" then recites the top three on her own list. "More than your brother?" her cousin asks, noting the conspicuous absence.

"We don't count," Rahel replies. It's a way of saying the opposite: that she and her brother count so much for each other that they're outside the whole game, outside every game available to singletons, forever and ever, amen.

Not all twins feel this way, but some do, and the way they live their lives reflects that throughout adulthood. They talk on the phone every day, they have regular twin-only vacations, they buy houses next door to each other, they co-parent each other's kids. In rarer cases, they continue to wear identical outfits way into their golden years or sign up for twin-to-twin dating agencies, with the goal of marrying a matching pair. Some studies suggest that identical twins are less likely to marry than singletons are: a quarter of monozygotic males and nearly a half of monozygotic females stay single for life. *Wrong!* society scolds. *Wrong wrong wrong wrong wrong!*

Can monogamous romance really be as overwhelmingly superior to all other human bonds as this ideal claims? Clearly, many marriages are conflict-ridden, abusive, or soul deadening. Many friendships, on the other hand, are highly supportive, fulfilling, and committed. To take one striking example, Deena Lilygren writes of her platonic friend Maggie in the article "I Love Living with My Best Friend, So I Bought a House with Her and Her Husband," "Our ability to cooperate and coordinate projects together is unprecedented in my life—I felt like I had a partner in life that I hadn't had before, not even in my marriage."

If we wanted to try out a less amatonormative world, what would that look like? Brake suggests we narrow the scope of marriage rights (for instance, tax breaks, hospital visiting privileges, housing preferences, and Social Security and health insurance benefits) to those relevant to caregiving, then extend those to all who provide sustained care to loved ones. Who cares for your welfare, she argues, is more important than who you're sleeping with and is a more appropriate focus for state attention, too. Beyond that, she says, we need to stop insisting that amorous dyads of the traditional type are essential to a meaningful and ethical life, and marginalizing, denigrating, and demonizing those who aren't part of one. Why not get our social ideals more in line with how many of us now actually live—in varied, creative, flexible networks of human relationships that may or may not involve any combination of romance, sexual exclusivity, cohabitation, child-rearing, shared finances, and biological ties?

We might also say that in a non-amatonormative world twinship would neither be romanticized nor pathologized. I admit I tend in the first direction, so I find it refreshing to read about real-life twins who aren't particularly close or who hate each other's guts. Tegan Quin, of the band Tegan and Sara, says that, despite her and her twin's success in working with each other, "we have very little access to each other's interior world." When she and Sara were kids, she writes, "Our shared best friends acted as a conduit between us: we confessed to them what we couldn't tell each other, and knew they'd pass along the message." In 2017 a pair of twin *Playboy* models were arrested after "Kristina punched Karissa in the face and Karissa threw an ADT security monitor at her sister, knocking out her teeth." In 1996, Jeena Han of Irvine, California, attempted to murder Sunny, her twin and high school co-valedictorian. I surprised myself recently when I came across a 1986 diary entry in my seven-year-old scrawl with a "list of hard things" that included "trying not to shout at my sister, when she does something I don't want her to I feel like exploding." Maybe things weren't *quite* as idyllic in Julia's and my childhood as sentimental me likes to remember them being.

But the idea that all twinships tend toward mental illness, violence, and premature death is absurd. Real-life twins don't generally "turn out to be mentally defective, low achievers, solitary loonies, criminals, and truants," as the twin researcher Alessandra Piontelli reports their parents fearing. If anything, adult twins seem, on average, marginally happier and healthier than the standard human. They're at lower risk of depression and suicide than singletons, and there's evidence they're underrepresented in psychiatric populations more generally. They also have higher life expectancy, once they get past their vulnerable first year. The extra years are especially marked in male twins, who, researchers suggest, may engage in less risk-taking than their singleton peers for fear of hurting their twin were they to injure themselves. ("For the love of Zeus!" you can hear these latter-day Polluxes cursing to their Castors. "Step *away* from that chariot, bro!") The one form of mental distress in which twins do have a significant lead over singletons is grief, when one twin dies. It'd take a hard-core investment in amatonormativity to deduce from this that their relationships are *bad*.

• •

I recently spent a sabbatical year in New Zealand, living with my parents in the seaside bungalow where Julia and I grew up from age eight to twenty-one. A couple of years before, Julia had returned from Luxembourg and built her own house in our parents' backyard, so she and her daughter could enjoy living with an extended family after her divorce. Julia's new house is what you could more or less accurately describe as "a modest rectangular modernist box, raised off the lawn, wood paneled inside, except for floor-to-ceiling windows at the front with a wide view of the coastal sky." We spent a lot of time in it that year, having cups of tea together and baking scones.

Julia, who is worse at repressing things than I am, worried repeatedly that we'd both feel awful when I went back to America. Hanging out all the time for a whole twelve months, on the exact geographical site of our happy childhood together, faking early retirement at

Camilla's, coming full circle at the age of forty-two—weren't we setting ourselves up for psychological meltdown on parting?

"I'm gonna miss you so much, Lena!" Julia would wail.

"Yeah, I know," I'd reply, then instantly change the subject, as if I had a stone for a heart.

As it turned out, I was distraught for six weeks after my return, whereas Julia was elated. Shortly after I left New Zealand she and a nonbinary friend she'd been hanging out with confessed their love to each other and started dating.

"I feel like it's not a total coincidence this happened 48 hours after your departure," Julia texted me brightly while I wept with homesickness on my bed in Boston. "It's interesting that I slotted them in the week after you left and we do things very similarly to you and me—like sharing all our meals at cafés and making each other cups of tea and working together."

I could have been offended, I guess, but instead we both thought it was hilarious. We reminisced about how Julia had done this before, when I moved to Australia after college and she started spending all her time with her new best friend, who wore her straight dark hair long, as I did, and who everyone assumed was her sister. My response to Julia's absence, all the way through my twenties, was the polar opposite: I avoided having any female friends at all. I didn't spend much time examining the motivation for this, but when I did, I told myself it was because I needed ample personal space while I grew into my new identity in America.

"Maybe you should get a female best friend or something?" Julia suggested over the phone when I reported feeling lonely.

"Women are always in your business!" I replied. "They can tell when you're feeling bad and then they ask you to talk about it!"

"Right?" Julia said. "Gross."

"Anyway, I don't even know how to do it anymore," I whined. "It's been years. How do you make friends with a woman?"

"Two things," Julia said matter-of-factly. "Compliment them on something—like their hair or sweater or whatever. Then share something personal about yourself, so you seem open and vulnerable. In that order."

I wanted to say, *That's so gendered!* But my ruling out all women as intimate friends on the basis of their gender alone made that a tricky rhetorical move for me. I tried out Julia's advice at a party the following week and it was incredibly effective.

Do I feel threatened by Julia's new person, my latter-day replacement? I don't, and when I ask myself why, it doesn't feel as if it's because I assume, as her twin, I've already won. It just seems easier for me not to view the situation as competitive than it did with Julia's past romantic partners. I'm tempted by the thought that this has something to do with all three of us not being straight. Sure, queers are as rivalrous on average as the rest of our scrappy species, but the specific idea of twins and their siblings' partners being sworn rivals in love reads as pretty heteronormative to me these days. It reminds me of the jealousy some men report when watching their wife fall in love with their new baby. (As Arthur Huntingdon puts it in Anne Brontë's *The Tenant of Wildfell Hall*, "That's more, in one minute, lavished on that little, senseless oyster, than you have given me these three weeks past.") We're meant to be empathetic about this: major life transitions are hard, relationships are complex, men have feelings, too, et cetera. But I confess that what I tend to think is, *Why not share her, asshole?*

I was arguably a bit of an asshole myself in the past when it came to Julia. It didn't show up in my behavior because it didn't need to: I never felt under any genuine threat. But it was there in my attitudes: my mythic conviction that twins across the board were always a better bet than romantic partners, my belief that I had to carve out a number one spot in my heart for Julia and that she had to do the same for me. What's this obsession with rationing and ranking your intimacies? I want to ask my younger self. Who said you have to convert everyone you love into horses and force yourself to pick a winner? I know who said it, actually, and I'm trying not to listen anymore.

Though I still like the idea of twin retirement at Camilla's, the vision feels different to me from how it used to, less defensive, more open-ended. Maybe Julia and I could live at Camilla's *with* Julia's new partner or their successor, I find myself thinking, and whoever I do or don't end up with? Then I remember my need for "personal space" and back up a little. How about our living in *adjacent* houses, I think, as Julia does now in our parents' backyard, in a kind of family compound? And then how about more houses for my friends and whoever they love, and their friends, and theirs—and I see my vision of a more expansive way of living turning from a two-person refuge into a world that fits all of us. I'm temperamentally allergic to utopias, but I guess I can sign up for one if it starts innocently with Julia and me, arthritic and in cardigans, in a vintage house by the sea. You can back into a revolution this way, as we all can when we run into a pair of twins and allow ourselves to see their relationship for what it is, rather than for what it resembles, or for what we desire and fear it to be.

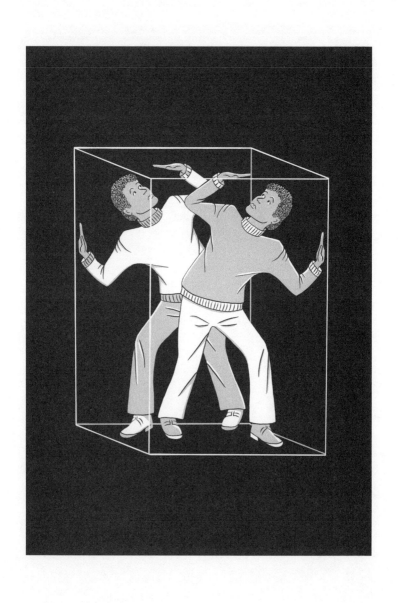

4

HOW FREE ARE YOU?

In January 2021 I was living with my parents and Julia in New Zealand, as a temporary escapee from the United States, which hadn't been performing at its personal best for the past year. I hadn't slept well in months, I felt edgy throughout the day, and I spent a lot of time wondering about the security of my possessions, which were scattered across two countries, three cities, and a set of basements and cars. ("I'm between apartments," I told people evasively, like someone who'd been evicted, dumped, or fired.) But I pride myself on being a cool customer—Julia: "LOL"—so I told myself I was coping just fine. Then the January 6 insurrection happened.

"My eyelid feels weird," I told the woman at the drugstore the next morning. "The right one—it's red and itchy and flaky."

"Have you had this before?" she inquired.

"Nope," I said.

"Have you been experiencing any stress lately?" she asked.

"Well, I'm an American citizen," I said.

"Okay, we're definitely giving you this steroid cream, then," she replied.

I told Julia about this exchange when I got back to the house.

"Wait, what?" she said.

"You know, it's funny," I said, "like—"

"I *know*," she replied. "But look at my eye."

She pulled down her right eyelid and there it was: an identical rash in the identical place on her identical eye.

"Have you had this before?" I asked.

"Nope," she said.

We blinked, identically.

No one else in our household or social circle shared our problem, so we ruled out the possibility that a new eyelid pathogen was gearing up to halt what was left of human society. The most likely explanation was that I'd caught something less virulent and passed it on to Julia. But I couldn't dismiss the more thrilling thought that we'd each independently developed our affliction, on the shared biological basis supplied by the single egg that formed us way back in 1978. What if something had been lying inert in Julia's and my genome for over four decades, I marveled, just waiting quietly to manifest itself in our two bodies exactly 15,369 days after our joint birth? Like those cicadas that lie underground for precisely seventeen years, then bust out, fuck, and die, maybe this eyelid rash had been saving itself up, holding back for just the right epidemiological/political conditions, undisturbed by the multiple divergent events that Julia and I had experienced since exiting the womb. Julia learns five languages, moves to Luxembourg for ten years, gives birth to a child, returns to New Zealand, takes up aqua jogging, reignites her passion for drawing . . . eyelid rash proceeding as planned! I major in philosophy, move back and forth between Boston and San Francisco, buy myself a ukulele, write three books, embark on recreational Ping-Pong . . . eyelid rash still on course, bitches!

"Isn't that wild?" I asked a woman I was on a date with a month or so later. "For it to happen the first time ever, the exact same week, on the exact same eye? For it to disappear, maybe forever, in the exact same four days, after the exact same number of hydrocortisone applications? It's like a compressed eye-rash version of Julia falling in love with a woman for the first time in her midthirties, and then me doing it two years later."

"Really?" my date gasped. "That happened?" She munched on a fry, meditatively. "It doesn't make you feel very free, does it?"

Tales of identical twins who undergo similar experiences often produce this reaction. "Twins cast the existence of free will into doubt," someone intones, when the latest story about identical twins separated at birth appears in the news. Or "twins raise deep questions about human freedom," we're told, in a puff piece for the latest twin novel or film. Hearing this kind of thing, I've tended to nod along gravely with everyone else, then get on with my business. But it occurred to me suddenly that it'd never really been clear to me what people meant when they said these things or what reasons they had for saying them. I found Julia's and my tandem eye rashes and late-in-life queer conversions remarkable, sure, but neither of them had made me feel *unfree*. It also struck me as odd that I've never been much moved by these allegedly urgent doubts about freedom, given that my people—Team Twin—are widely thought to raise them in their most acute form. Of all the philosophical topics I've studied, free will is way up there with those that have left me the coldest. If people like Julia and me represent a major threat to human free agency, I found myself wondering mid-date (a professional hazard), why, again? And why don't I seem to care?

· ·

If you ask people why twins make them doubt their own freedom, they're most likely to point to the results of twin studies. In 1875, Charles Darwin's cousin Francis Galton noted the scientific usefulness of single-egg twins being genetically identical. It follows, Galton pointed out, that any differences between a pair of them can't be explained by differences in their genes, but must be due to variation in their environments. The classic twin scientific method exploits this nifty situation in a variety of ways. One is to compare data on single-egg twins who were separated at birth and raised apart. If scientists can show that single-egg twins raised in different environments are often or always similar to their twin with respect to a given trait, that suggests a strong genetic influence on the trait. Another variation brings different-egg twins into the picture. Different-egg twins share only 50 percent of their genes, no more than a standard pair of singleton siblings. If scientists can show that, on average, single-egg twins raised together are more similar to each other with respect to a given trait than different-egg twins raised together, that again implies that the trait is strongly genetically driven.

Thomas Bouchard, who ran the famous Minnesota Study of Twins Reared Apart in the 1980s, has called the twin method "the Rosetta Stone of behavioral genetics," allowing scientists a godlike reach into the previously hidden roots of human features and behavior. Criminality, alcoholism, smoking, schizophrenia, academic success, job satisfaction, physical prowess, anxiety, divorce—thanks to twins and their observers, scientists now have data-driven "heritability" estimates for all of these and more.

One thing they've discovered, famously, is the existence of some truly wild similarities in twins separated at birth. Bouchard's most well-known pair, "the Jim twins," found on being reunited at thirty-nine that they'd each been christened Jim, called their pet dog Toy, married a Linda then a Betty, and named their son James Alan / James Allan. They both worked security jobs, drove Chevys, spent their weekends doing woodwork in their garages, and vacationed at the same beach in Florida. The results of their personality, intelligence, and brain-wave tests came out almost identical; they had debilitating headaches; they consumed large amounts of nicotine and beer; they weighed exactly 180 pounds, ground their teeth at night, and bit their nails to the quick during the day. Other reunited twins in Bouchard's study were found to each leave love notes to their wives around the house, compulsively fidget with rubber bands, sneeze in elevators as a prank, flush the toilet before using it, and walk into the ocean exclusively backward.

Twin studies comparing single-egg and different-egg twins have generated similar, if less showy, results. Some traits have been found to have low heritability: the religion a person goes for, their political affiliation, their sense of humor, and their food preferences. Environmental differences seem to largely explain the variation there. But the heritability of the traits that many of us care most about—personality, temperament, and intelligence—looks pretty high, at least in the Western societies where most twin studies have taken place. On what are known as the OCEAN personality factors—openness, conscientiousness, extraversion, agreeableness, and neuroticism—single-egg twins are much more similar to each other than different-egg twins are. The heritability of emotional intensity and degree of activity is also relatively high, as is, to a lesser degree, that of impulsivity, adaptability, distractibility, and dominance.

What might all of this have to do with freedom? Well, an appealing line of thought goes, if your genes strongly affect such insignificant features as your tendency to twang a rubber band around your wrist and such consequential features as how much of a party animal, drama queen, or control freak you are, it looks as if much of who you are and how you live is reducible to your DNA. Given that you can't change your DNA, how free an agent can you be?

Many took the results of early twin research to confirm Galton's doubts that "nurture can do anything at all, beyond giving instruction and professional training." But if you look at the actual numbers, it's clear this conclusion was overdrawn. Except for physical features such as eye color, it's rare for any trait to score higher than 60 percent on the heritability index—which isn't nothing, but suggests a major role for environmental factors, too. This caveat doesn't help much, though, if it's freedom we're worried about. When scientists refer to "environmental" determinants of human traits and behavior, they have a diverse bunch of things in mind. They can be referring to the family you're raised in, the peer group you have as a child, the culture you're exposed to, and the individual relationships you enjoy along the way. They can also have in mind nonsocial factors, such as the hormonal and nutritional conditions you experience in the womb, the trauma you undergo during birth, the viral illnesses or accidents you suffer, and the allergens floating around your neighborhood. We have just as little control over many of these things as we do over our genes.

In some cases we're *less* able to alter environmentally driven traits than genetically driven ones. Though your natural hair color is fully genetic, you can dye it another color in about ten minutes. You're unlikely to undo the effects of your PTSD in that time, if ever.

If the naturists are right, you're mainly the puppet of your genes; if the nurturists are right, you're mainly the puppet of everything else; either way, it looks as if your status as a puppet is secured.

• •

I first came across the debate over metaphysical freedom in a seminar I took in my senior year of college. The class was held in a garret on the second floor of the rickety cottage that housed the philosophy department. I and six other majors crowded into the room with the professor, a recent arrival who displayed what seemed like a distinctively American combination of earnestness and ebullience. He was short enough to not hit his head on the roof when standing, so he lectured there in front of us, wildly gesticulating, while the bright fire of rational argument beamed from his chest like a searchlight sweeping the dark, unsettled sea before him.

My memory of the details is fuzzy, but our first class no doubt tracked the standard intro to the subject, which runs like this:

"Most of us go about our lives assuming we have a significant degree of control over our values, choices, and actions," the professor begins. "Sure, we may run up against some everyday impediments—say, our micromanaging boss, our limited cash flow, a misogynistic landlord, or a racist cop. But we believe that how we respond to the world, including to those kinds of constraints, is largely up to us. In living our lives, we think, we're repeatedly faced with a range of different options for action, and we select and enact some of those options and reject others, using something we call our free will."

Everyone nods.

"That idea doesn't sit well, though," the professor continues, "with a basic assumption of modern science. Physicists have been claiming for centuries now that the universe is deterministic: every event that occurs is the product of the laws of nature combined with all prior events. If that's so, every event that happens could only have happened in the exact way that it did."

An example gets thrown out at this point, a cozy one designed to take the edge off the existential hellhole we're about to be dumped into.

"Consider, say, my order of poached eggs, instead of scrambled ones, for breakfast this morning. It may have seemed to me, as I hovered over the menu, that I could have ordered scrambled eggs instead. But, if determinism is right, my ordering poached eggs was always the only real option. In fact, my having poached eggs was settled thirteen billion years ago, when the Big Bang occurred and set the universe inexorably in motion. We may not be able to pinpoint all the steps in the causal chain from primordial explosion to breakfast, but that's because human knowledge is limited, not because the steps aren't there, stretching way back into the ultradistant past. A complete, mature science would be able to exhaustively reveal every interconnected force leading up to each of our actions, making clear that in every case, no alternative action was genuinely available to us."

At this juncture some smart-ass with a hard-science class under their belt raises their hand to object. Back in the days of Newtonian physics, they say, scientists believed in universal determinism. But, thanks to the twentieth-century discovery of quantum mechanics, physics now includes laws of nature that are fundamentally probabilistic. Those laws don't generate unique predictions; they allow for multiple outcomes. Prior events may make it more likely that one outcome rather than another takes place, but they don't necessitate that outcome. Sometimes a "quantum event" or "jump" intervenes, and a different outcome from the likeliest happens, for no detectable reason at all.

Then a philosophy bro sweeps in and cuts that other person off. "Let's say that quantum physics is right and the universe is indeterministic. When you're poised on the brink of action, then, whether you go for option A, poached, or option B, scrambled, is, at least sometimes, totally random—arbitrary, uncaused, and fundamentally inexplicable. The universe shudders or sparks at an invisible microscopic level, and option A is selected, but not due to anything that happened prior to that moment, or any general law. Okay. But how does that help? We still don't get free will; we still don't get to be in control of our actions, because if action A or action B happens randomly, by chance or luck, it's clearly not *us* who are deciding the outcome."

"You're getting it," the professor says encouragingly. "And notice that not just one, but two alarming conclusions are now beginning to look inevitable. The first is that free will is impossible. Human behavior is either predetermined or it isn't: there's no third option here. So if both of those routes imply that free will is impossible, that's what it is."

Pause.

"But things are worse than that." (This bit I do remember clearly—Americans are so dramatic!)

"The second conclusion is that the very idea of free will is incoherent. We want our actions to be neither predetermined nor random. But that doesn't make sense! If something's not predetermined, it just *is* random. So it looks like free will isn't just impossible; it's totally unintelligible."

• •

Several of my classmates exhibited a kind of panic about this dilemma. It seemed they'd been painted into a conceptual corner and had no idea how to get out. And if they couldn't get out of that corner, was there *any* corner they could get out of—you know, metaphysically speaking? If our professor was right, their previous picture of themselves as free agents faced with a wide array of options, any one of which they could pluck into actuality at will with their own powerful hands, was completely unsalvageable. Logic dictated that they couldn't do anything truly free at all! The result was acute existential claustrophobia.

I, in contrast, felt quite relaxed about it all. For the rest of the semester, as the various positions in the debate were rolled out, and the objections, counter-objections and counter-counter-objections were batted back and forth like a slow-motion game of conceptual tennis, my mind, I confess, would wander. The chirping birds, the paper for my French-poetry class, my lunch plans with Julia—pretty much anything was better at holding my attention than the apparently super-important philosophical question of whether humans are genuinely free agents. I slogged my way through the seminar, got my A through pure force of will, and, despite pursuing a successful career as a philosophy professor, thought about the subject for less than probably thirty seconds for the entire next twenty years—until my date made that comment about my eyelid.

Maybe my college self was of the "compatibilist" school. Compatibilists argue that we can have a perfectly coherent kind of freedom of action even if our behavior is determined by prior events. Free will doesn't require that you *could* have acted otherwise than you did, compatibilists say, all it requires is that you *would* have acted otherwise had you wanted to. If physical coercion, psychological manipulation, lack of opportunities, or physical or mental impairments play a role in your action, then, yes, your free will has been curtailed. But if those kinds of obstacles are absent and you're just doing what you want, you're a free agent. This is the way we talk about freedom when going about our actual lives, compatibilists suggest, and that intuition is right.

Another compatibilist way of putting it is to say that freedom only requires that *you* be the source of your actions, rather than something or someone else. If another human is moving your arm for you against your will, or blackmailing or drugging you, you obviously don't satisfy that source condition. But if you act on the basis of your own desires, your action springs from you. Sure, plenty of outside conditions and agents will have informed those desires. But your desires remain, right now, yours and no one else's. So in an intuitive sense you can be the origin of your actions, even if determinism is true.

This take is pretty appealing at first glance. The bad news is that it's vulnerable to a serious objection. Let's say, for instance, that I've got a bad habit of texting my toxic ex to ask if they want to hang out. I know I shouldn't text my ex, yes. I know I shouldn't *want* to text my ex, sure. I've spent the last two weeks telling all my friends that I *absolutely will not* text my ex again, ever, under any possible circumstances, yep. Then I . . . *OKAY, FINE, SO I TEXTED MY EX LAST NIGHT, WHAT'S IT TO YOU?!*

I satisfy the compatibilist criteria for free will. The source of my late-night text was me, and *if* I'd wanted not to send it, I wouldn't have. No one was there forcing me to. That means that this time, and each time, I contact my ex, compatibilists have to say I'm acting freely. The problem for compatibilism is that, when someone acts this way—compulsively, in ways they know they shouldn't—both they and their exasperated social circle tend to think of them as the very opposite of free. My friends and I all know that I won't want to stop texting my ex until I'm over their sorry ass. Given that I'm *not* over their sorry ass, I couldn't have done otherwise than text them last night. The compatibilist idea that I acted freely in sending that text because I *would* have resisted sending it if I'd *wanted* to sounds like a sick joke.

Compatibilists have made a lot of ingenious moves—including distinguishing between desires we do and don't identify with—to shore up their approach in light of this type of worry. But many people remain unsatisfied. Kant referred to the whole compatibilist program as "petty word-jugglery" and "a wretched subterfuge"—tell us how you really feel, K.!—and while college me did dally with it, I wrote a final paper siding against it because I couldn't see how it could be pulled off. Still, I left that class with the strong compatibilist-*ish* sense that, though determinism was true, I had as much free will as I cared about having. Recently I've been thinking that part of the reason for that is that I'm a twin.

∴

I care a lot about everyday forms of freedom: freedom from domination, oppression, or abuse, freedom to roam my local area or go on an international trip, freedom to choose my career or organize my afternoon. But the idea that we'd all be better off if we had the kind of *ultimate* freedom that people who angst about free will fixate on—where we indisputably control the causes of our actions, all the way down—that I don't relate to. These people who yearn for ultimate freedom, I think, like a man mystified by "women": What do they want?

The answer is complicated, but one deep thing they arguably want is to preserve their sense of self-worth, grounded—for those raised in Western cultures—in their status as self-governing agents. On this view, what makes us humans the elevated beings we are, unlike the benighted "lower" animals, or those in our own species we've unjustly dominated, is that we run our lives according to our own desires and principles, rather than being driven by mindless instincts, other people, or random forces outside our control. The philosopher Saul Smilansky claims that the idea that our actions are fundamentally products of external forces is therefore "extremely damaging to our view of ourselves, to our sense of achievement, worth, and self-respect."

I have some basic trouble getting my head around this. I just don't share the apparently widespread intuition that, were I to have ultimate control over my actions, I'd be more worthy of respect, more precious, and more dignified than I would be if I lacked it. Part of the problem here is that I can't see how such ultimate free will is even conceivable. It seems to require, as Nietzsche put it, that we be the causes of ourselves—that we have the power, "with more than Baron Munchhausen's audacity, to pull [ourselves] up into existence by the hair, out of the swamps of nothingness." For one thing, what would that even look like? (The philosopher Susan Wolf notes that "literal self-creation is not just empirically, but logically impossible.") For another, in what sense would this abstract self, who exists fully outside the causal nexus of genes and environment, actually be us? Who you are isn't separable from the biological and social aspects of your being: those things *are* you. This ghostly primal self-creator who doesn't have the eyes you inherited from your mom, or your dry British sense of humor, or the love for chinchillas you picked up from your first best friend—who are *they* to go around claiming that your freedom requires that you be created by them?

Beyond that, even if this valiant act of solo self-actualization were possible, I don't see what's so desirable about it. It sounds lonely, cramped, and dull to me. To wish that you'd got the whole job of self-formation done alone is to wish that you hadn't opened yourself up to other people, that you hadn't been curious about objects outside your skin, that you'd lacked the courage and hope to submit successive drafts of yourself to the universe for feedback. I like the idea that other people have helped make me who I am, the sense of complicity I feel when I notice my face, tone, or words mimicking those of someone who's been important to me. If I've folded them into my personal batter and baked them in, I can carry them round with me—like mini-Julias—when they've moved out of my orbit and are gone.

What I care about is how I am, not how I got here. I'm pleased with my features and achievements not because I made them, let alone made them all by myself, but because of what they allow me to do and enjoy. I'd suggest that, when we think about it, most of us view ourselves and other people this way. We respect and admire one another for our kindness, intelligence, sassiness, and beauty, not for the ultimate source of any of those things. If that's so, the Western commitment to extreme self-government doesn't line up well with what, deep down, many of us seem to value.

Sometimes I trace this take to my being a woman, or someone with a disability. Being totally in charge of your destiny, independent of the influence of others, seems like a macho, able-bodied thing to want, the result of the kind of lifelong privilege that allows you to forget that we're all vulnerable, dependent beings whose allegedly free actions are highly conditioned by social and physical forces. But I also think my attitude springs from the perspective of a twin (or at least a twin whose experience of twinship has been like mine—close and mostly happy). Such twins are less likely to be invested in individualism, and more inclined to value than fear the sort of intimate connections with another human that deeply influence who one is. As a result, they're less likely to locate their dignity in the impossible feat of having formed themselves, unaided, from scratch.

You don't have to be a twin to take this compatibilist line, and I'm sure plenty of twins are in the other camp. But I want to say that if there's a twin-like position in the free will debate, it's this one. Identical twins are likely to accept determinism, having seen its obvious stamp on their and their twin's lives from day one. And those who get on well with their twin are less likely than the average person to yearn for a personal causal history that features only themself as the star. I don't claim to have done a scientifically respectable study on this, but I've asked around. "How do you feel about the question of whether humans have free will?" I've quizzed twins I know or run into: students, my hairdresser, a random dude on a rooftop, my editor. "Bored," they reply. "Don't care," they say, shrugging. "Somehow that question isn't as compelling to me as one would think!" I didn't even bother asking Julia. I knew the answer already, and when I told her I was writing this chapter, the expression she generously struggled to dismiss from her face amply confirmed it.

Whether or not this generalization is true, I like the idea of it. We twins are so often used as avatars for the horrifying thought that humans are mere puppets of the universe. It'd be amusing if, instead, our experience suggested the opposite. While we're meant to throw you into existential crisis, I think we should actually make you feel better about your cosmic situation. Twins are best seen not as posing a threat to free will, but as providing reassurance that the kind of everyday freedom we all have is more than enough.

• •

I'm sounding very chill here, which, as noted, is not my general vibe. I do think we should be less caught up than many of us are with being ultimately free agents, for the reasons I've just given. So stories of twins separated at birth don't make me feel like a metaphysical doormat or trigger existential claustrophobia in me. But some of those stories do mightily unsettle me, for an almost opposite reason. Rather than raising the suffocating thought that things couldn't have been different, some separated-at-birth narratives make it obvious that things very much *could* have been different for each of us, even if in actual fact they aren't. What these twins throw disturbingly into relief isn't our lack of control over life, but the arbitrariness of our convictions.

About as striking an example as you can get is the real-life case of Oskar Stohr and Jack Yufe, a pair of identical twins who were separated soon after their birth in 1933. Jack stayed in Trinidad, where he was raised as a liberal Jew by the twins' father, while Oskar moved to Germany with the twins' mother, was raised as a Catholic, and . . . joined the Hitler Youth movement. That the presumably random fact of which parent you get in the divorce can determine whether you become a progressive Jew or a Nazi is pretty alarming. Oskar and Jack were understandably creeped out by it, but the rest of us can be, too, since their case vividly underscores the possibility that each of our values is determined by pure luck.

Most of us like to think of ourselves as selecting our beliefs and principles based on judicious reasoning, so the idea that we're just mindless recipients of our treasured credos makes us look foolish. The real worry here, though, isn't that we haven't chosen our beliefs and values, that cognitively we're undignified sheep. The bigger concern is about our beliefs and values themselves. If we haven't come to endorse them based on the reasons in their favor, how can we be confident that we've ended up with the right ones?

Amia Srinivasan calls this worry "genealogical anxiety": the fear "that the causal origins of our representations, once revealed, will somehow undermine, destabilize, or cast doubt on the legitimacy or standing of those representations." She underscores the phenomenon with a twin-like metaphor that was all too literal for Oskar and Jack: "What reason do I have for thinking that my beliefs are true, or that my values are genuinely valuable, or that my concepts grasp the contours of reality, if I could so easily have held contrary beliefs or values, or cut up the world in terms of rival concepts? . . . What am I to do with this other me, this shadow me, this me who believes the opposite of much of what I believe, who values what I disvalue, and who articulates the world in terms of concepts that are alien to my own? What if she is the right one, and I am the shadow?"

As Srinivasan notes, twentieth-century philosophy has a tradition of squishing this kind of anxiety. According to what's known as the genetic fallacy, it's a mistake to take the origin of your belief as a reason to doubt its truth. Let's say you believe that Harry Styles wrote "Watermelon Sugar" solely because of a random vision you had while high. Given that Harry Styles did write "Watermelon Sugar," who cares how you got there? But, even if in principle the source of a belief needn't discredit it, it's at least worrying if you have no independent reason to think the mechanism that gets you to a belief tends toward the truth. Drugs, genes, parental indoctrination, and culture are like that: they're unreliable routes to genuine knowledge.

How can you defend your convictions against the charge of arbitrariness if you know you've arrived at them in these unreliable ways? You could argue that you're "genealogically lucky" (to use Srinivasan's term). You could insist that you just happened to get the set of beliefs that latch on to reality—say, the belief that Nazism is wrong—and your unlucky shadow self didn't. But in doing this you risk sounding like a wishful thinker and a dogmatist, and those who disagree with your views are unlikely to be convinced. "I'm lucky!" anti-racist Jack yells. "No, *I* am!" antisemitic Oskar yells back. It's a stalemate.

This bothers me a lot more than the concern about ultimate control I've been talking about so far. Like any of us, I have a strongly held set of beliefs that are fixed points in my moral scheme. Some of mine are that society is unjustly stacked against racial minorities, women, people with disabilities, fat people, and queers, that capitalism has fucked us all over, that a climate apocalypse is coming, that God doesn't exist, that art can save your life, that cocktails and premarital sex are a great way to spend a Saturday morning, and that we should all vaxx up, abandon plastic, and become vegans. It's not a super-surprising worldview to come across in an East Coast liberal arts college professor who grew up in a middle-class white New Zealand family with a union organizer dad and a hippie artsy mom. Maybe as a philosopher I'm marginally more reflective than the average person, but I'm aware I didn't get to these beliefs via pure reason in a social void, even if I fine-tuned them a bit in grad school afterward. So I can pretty easily get into the mindset of my shadow self—an anti-tax, carnivorous, brunch-skeptical

Christian, say—who believes my convictions are completely unjustified. I badly want to be able to defend everything I care about in response to that kind of challenge. But can I? Can any of us?

Maybe not, but as someone whose job it is to talk with others about our reasons for our beliefs and values, I can't help but feel a little hopeful here. Most of my students arrive at college as dyed-in-the-wool progressives and leave it with those credentials intact or amped up. But in my colleagues' and my philosophy classes I've also seen omnivores ditch meat, anti-affirmative-actioners become anti-racists, believers divorce morality from religion, and soak-the-rich leftists become tax-shy libertarians. This suggests, at least anecdotally, that people are capable of changing fundamental beliefs they've had for a long time and care a lot about, in response to reasons offered on the other side. In those cases, the explanation for the values people end up with isn't purely a matter of arbitrary nonrational factors, such as their parents just happening to hook up during a snowstorm in Peoria on March 18, 2003. Reasons and reasoning are a significant part of the causal chain.

Reflecting on where our convictions have come from can be an important part of this kind of evolution, in motivating us to stand back from the beliefs and principles we've inherited and examine their foundations afresh. If we ask ourselves, "What reason do I have to be a feminist, other than the fact that my mother was one?"

the feminism or anti-feminism we end up with will be a position we've actually considered the rationale for, not just one we've inherited, and to that degree, one we have greater reason to trust. We can never be sure after doing this that we're not kidding ourselves that the beliefs and values we end up with are the ones it's independently reasonable to adopt, rather than the lingering effects of our socialization. But we're doing the best we can within the limits of the human condition.

To the extent that stories such as Oskar and Jack's motivate this kind of reflection, they can justifiably make us feel more, rather than less, secure in our convictions. So this is another case, I want to say, where we should see twins not as shoving us into a metaphysical abyss, but as offering us a hand (or four) in getting out of it.

• •

Stories of twins separated at birth sometimes inspire existential claustrophobia, sometimes moral panic. Other times, they bring on a feeling akin to grief. This comes up in *Identical Strangers*, a memoir co-written by twins Elyse Schein and Paula Bernstein. The sisters were thirty-five when they learned of each other's existence, the age Jung claimed is the human average for a midlife crisis. As they tell it, the surprise discovery of their twinship sends both Elyse and Paula into instant meltdown.

Paula is mainly unsettled by the challenge Elyse poses to her sense of herself as a distinctive individual. "Part of

my reluctance about having a twin," Paula writes, "is due to my steadfast belief that I am . . . a true original. Whenever people comment that I remind them of someone they know, I grimace." Speaking to her sister on the phone, she finds herself preoccupied with "weeding out which elements of [her] personality are uniquely [hers]."

Elyse isn't worried about that. She's confident going in that she's the freewheeling artist of the pair and is troubled instead by the vivid real-life counterfactual Paula's life represents. "The vision that scares me the most is that she has conquered her solitude and has settled down with a soul mate and a child," Elyse writes when gearing up for their first meeting. "If I witness her domestic bliss, will I regret having opted for a liberated but solitary existence?"

I'm Elyse in the Schein/Bernstein twinship, at least in this department. Though I was married for a while, I never had or wanted kids, and I've now lived solo for years. I do more or less what I want, spend most of my money on myself, and set off on trips as carefree as my constitution allows, with only my relatively self-sufficient cat to make arrangements for. Julia's been more stably partnered throughout her life, and she's raised a kid for over a decade now. More generally, her social circle is packed with people, many of them in varying states of need, who she feels responsible for. She moves toward social connection like it's a fire in a cabin on a cold night, and when she gets there, she stokes the flames higher. I sit happily fireside, too, but I'm always worried it'll get too hot. I have one eye looking out the window, wondering what's over that snowy hill.

This is the only real source of tension between Julia and me because, though I love her more than anyone else, she's one of the people I try to get away from. She's used to it and doesn't take offense, but sometimes it hurts her. I know I'm hurting her, but I do it anyway, and that hurts me, too.

This used to be a supersensitive subject between us, only rarely approachable in conversation, but we've recently sublimated it into a kind of joke. When we watched Disney's *Frozen* with Julia's daughter, we saw our central sisterly conflict cartooned on the screen. In case you haven't hung out with a kid in the past decade, here's how the midsection of the movie goes. Princess Anna responds to the news that the castle is having its first party in a decade as if she's just had a shot of heroin to the arm. Her sister, Elsa, charmingly works the ballroom for a while, then abruptly escapes to a distant ice palace, vowing never to return. Anna instantly follows her out there, with a team of new pals, to drag her back to civilization. Elsa is legitimately worried she'll freeze the whole village, due to a special power she can't control, but she's also low-key high on the radical freedom of being a literal ice queen. So, though she loves her sister more than anyone else, she refuses to go back. When Anna refuses in turn to leave, Elsa commissions a giant snow monster to get Anna the hell off her property.

These days, when I'm trying to do something similar to that, Julia begins to trill "Do You Want to Build a Snowman?," the song toddler Anna sings when Elsa locks the door to her own room. Then I trill back the first few lines of "Let It Go," the anthem Elsa blasts when she tries to ditch the human race forever. It releases a little tension on both sides.

Most of the time I'm content to be Elsa, but sometimes, especially when I'm leaving New Zealand to go back to America, I feel as if I've driven a shard of ice into my own goddamn heart. I'm free, I tell myself, but for what? I'm leaving the cozy family compound—our childhood home at the front, with our lovely parents still in it, Julia's snug home nestled behind it, with her cuddly daughter installed—for my empty residence in the chilly north. I *chose* this, I remind myself as the plane lifts off—for the sake of my education, my career. It didn't have to be this way. I could have had Julia's life. It would look just like that.

You don't have to be a twin to feel this kind of existential grief, or to have it inflamed by the sight or idea of twins. Freud attributed our collective cultural fascination with the idea of a double to "all those unfulfilled but possible futures to which we still like to cling in phantasy, all those strivings of the ego which adverse external circumstances have crushed." Albert Goldbarth's poem "Alien Tongue" imagines a language that includes a wistful tense with the meaning "I would have, if I'd been my twin."

Is there a way for any of us to feel better about our lost phantasy futures? One thing we can do is tell ourselves we're not irrational to feel the way we do. In his book *Midlife*, a philosopher's guide to getting through a midlife crisis, Kieran Setiya writes that it makes perfect sense to feel a sense of loss when considering the road untaken, even if you're happy with how your life has turned out and don't see any of your major life decisions as mistakes. That's because most of our big choices in life involve options with incommensurable value. The alternatives we face are good in very different ways, which means that the gains we get from taking one can't simply cancel out the losses we suffer from not taking the other. If the joys of a free-wheeling single life are incommensurable with the joys of partnered-and-parental life, neither can fully compensate for the other: both inevitably involve loss that can't be redeemed. So even in the best-case scenario, where you endorse all your past choices and feel good about where you've ended up, it makes sense to feel sadness—maybe even searing longing—for what you've missed out on.

That human life inevitably involves uncompensated loss sounds like bad news, but there's a positive way of looking at it, too. The grief we feel for lives unchosen is the price we pay for a world that's overflowing with many and diverse kinds of value. It springs, Setiya writes, from "the fact that there are so many different things worth wanting, worth caring about, worth striving and fighting for, too many ever to exhaust." The only way to avoid this grief would be to live in a world that was less rich, or to be someone who was incapable of appreciating the richness there is, a person who loved only one thing. Who would want that?

So Elyse could have reassured herself, before meeting Paula, that any regret she felt as a result would just be the cost of living in a world that abounds with goodness and beauty. But the meeting would still be risky, and the devil would be in the details. Even if it's clear to you that becoming, say, a childless writer was on balance a mistake,

Setiya suggests, it's rational to keep that alternative vision cloudy and abstract. There will usually be plenty to value in the activities and relationships you *did* choose, and focusing on the particularities of those is a handy way to avoid getting swamped by regret. "It is rational to respond more strongly to the facts that make something good, in all their specificity, than to the featureless, generic fact that something else is better"; "in order to avoid regret, you must preserve a measure of oblivion."

It's all very well for a singleton to say that. Setiya writes, "An actual life is not a thought experiment; a sister is not a counterfactual self." But a sister who's a twin comes pretty close to being that. As an identical twin, you can *actually find out* the answer to the behavioral geneticist Kathryn Paige Harden's question, "If you rewound the tape and started anew from the exact same genetic and environmental starting point, how differently could your life go?" When Elyse meets Paula, that's exactly what she's going to discover, in maybe excruciating detail. It's what I discover every time I hang out with Julia.

This is one way in which being a twin is genuinely worse than being a standard human being. Some twins experience acute or chronic envy and jealousy toward each other; in the really bad cases, their relationship is torn apart by it. I wouldn't want to underplay the reality or pain of that experience, but that kind of twinship dysfunction gets plenty of cultural airtime. So it's worth pointing out that, if existential regret is likely to be more acute for twins, the sense of richness that Setiya talks about can be enhanced for us, too.

Watching Julia dart off in directions different from those I've taken has often made me feel vividly what a vast array of paths are open to all of us collectively, whether the outcome for each of us is determined or not. That vision of multiple possible lives gives me a sense of expansiveness—I want to say it feels *freeing*—even if I'm not convinced that ultimate free will is a thing. Who could have told, back when Julia's and my egg split, what varied adventures awaited us both? Whether those adventures were preloaded into the universe, whether we're fundamentally responsible for them, or not, how interesting and wonderful to live them out. And what a cosmic privilege to see your particular genetic and familial plot play out not just once, but twice. If you're an identical twin, you can have, in effect, both the poached and the scrambled eggs. *I would have*, you can say over and over, *omg, I would have!—if only I'd been my twin*.

5

WHAT ARE YOU FOR?

I was sitting in my friend Sarah's living room a couple of years ago, chatting over snacks, when she mentioned that she'd "had a lot of social power in high school." I had sort of suspected this, but she hadn't stated it directly before, and I wondered how exactly it had shown up.

"What kind of social power?" I asked.

"I was beautiful," Sarah said matter-of-factly, "and much desired."

"Right," I nodded, as if this were a normal thing to say, and not something I couldn't imagine claiming about any period of my life ever, but especially that one. I crunched down faux nonchalantly on a cracker and we moved on, but her announcement was hard to get out of my head.

I knew Sarah had dated a string of acid-dropping musicians from her teens on, had led a wild youth in New York, Paris, and Los Angeles, and had posed for the covers of a series of erotic novels to get extra cash in college. In her forties, she still had curly red hair, a porcelain complexion, and a great figure, and she seemed fully at ease in her body: it moved fluidly in and out of chairs and rooms at her command, and she stretched it languorously, panther-style, while staring you straight in the face like, "What?" Recently, while imitating her cat pouncing on a toy, she had shaken her ass comically and it had looked kind of hot, whereas on me it would not have, and also I would have put my back out. Of course Sarah had been wanted in high school. It was totally on brand.

My high school years were a continuation of my own brand with respect to sex and seduction, which was to be super awkward about it, and then awkward about being awkward about it, and then possibly even awkward about that. This had been my vibe ever since my first discovery of sex, some rainy afternoon in my sixth or seventh year. I was hanging out that day in the basement of my friend Clare's house when she asked if I wanted to see something. When I said yes, she took a padded photo album off the lowest shelf of the bookcase, laid it out carefully in front of us, and flipped the cover open. I'm not sure what I was expecting, but it definitely wasn't a studio-quality portrait of Clare's parents, lying on the shag pile carpet on which their daughter and I currently knelt, their limbs entwined, in boudoir lighting, with absolutely nothing on. The only forms of concealment were the father's giant mustache and the mother's superlong and straight hippie hair, neither of which succeeded in reaching the lower portions of their respective bodies, though a valiant effort was made. Most of the similar pictures on the following pages were in color, some in black and white, with a solo portrait here and there. A few pages in, the dogs got involved. The family's two golden retrievers, equally hairy and louche, stretched themselves out against their owners' naked bodies, merging everything into one indistinguishable seventies-brown blur of carpet, flesh, and fur.

When I thought about this primal moment later, what mainly struck me was Clare's attitude to her parents' tenderly preserved soft porn shoot. Who knows what was really going on in her mind, but to my junior eyes she simply seemed fascinated. For my part, I was horrified, and highly motivated never to look at that album again. I felt judgy about it, too. I was pretty confident that if I'd found my own mother and father posing naked with the cats, I wouldn't have been broadcasting it to the whole neighborhood.

As we moved into our tweens, my school friends all seemed eager to develop their knowledge and power in relation to sex, whereas I preferred to think and talk about literally anything else. My first best friend lived at the end of a street perched on a massive bluff with vertiginous views of the ocean. Her house seemed on the point of falling off the edge of the world, especially when the frequent coastal gales whipped through, shaking every window on the main floor. On our after-school dates, we'd have a snack and then retreat to the bunker of the basement to play with Eliza's Barbies. Eliza would always start off by announcing that her mom forbade her to remove Ken's pants. Then she'd look me directly in the eye and gleefully rip off his jeans. It was totally obscene, it was so wrong, each time I couldn't believe she'd do it! While Eliza waved Ken's missing genitals in the air and made lewd scissor motions with his thighs as the house shuddered above us, I'd concentrate superhard on assembling a vacation outfit for Skipper, till my companion got bored and moved on.

My next best friend, in middle school, was a future graphic designer, and we spent most of our time together drawing crazy lineups of people and animals in bespoke outfits, on pieces of paper we swapped back and forth in class. But one day, when we were down in Katie's basement with a group of other girls, the agenda abruptly shifted. Suddenly Katie was leading a pole-dancing class, centered on the column that supported the two upper floors of her house, along with her entire oblivious family. As I watched her and the other girls wrap their legs around the pole, arch their backs, and sink dramatically to the floor, I was overcome by alienation and shame. Theoretically I could join them: it was just a matter of grabbing the pole next and imitating what I'd seen. But it was obvious to me that my doing so would be a disaster. I'd look ridiculous, I'd mess it up, and it'd be apparent to everyone, what I'd felt inchoately for years: that whatever *it* was, they had it and I never would.

My tween aversion to sex mainly reads to adult me like that: a fear of having my incompetence at it exposed. I had a major musculoskeletal disorder, and I'd been raised by my parents and doctors to see my body as a serious problem. I didn't think of myself as unappealing on the surface—between injuries, I looked more or less like my friends—but I knew that, as a structural matter, every cell in me was defective. Against that background, the idea that I could exert power over my body to exert power over boys and their bodies would have been straightforwardly implausible to me. *What* power over my body? I could fake it, but the clock would be ticking and soon enough the game would be up. I suspect I saw it as better not to enter the game to begin with, given that it could only end one humiliating way.

I was also, maybe, a baby feminist of the sex-negative type, without knowing there were words for that. As a woman, staking your worth on others' desires reliably sets you up for harm and debasement. Junior me may have intuited that and felt the safer route to self-respect was to step outside the whole looming arrangement, by avoiding the sex part on which it seemed to hinge. Perhaps I'd have ditched this strategy by becoming super sex-positive about some particular person in high school, had there been any viable candidates around. But sex was still way off the agenda for me then. Julia and I chose to attend a girls-only school, mainly because there'd be no teenage boys there to rush through the corridors and knock us over. The idea that we might direct our romantic attention toward our fellow students was so foreign that it never occurred to us. We had what we now recognize as a joint mega-crush on one of them, an older student who was beautiful, talented, and aloof. We schemed together about how we could spend more time with her, joined the club she led so we could stare at her without seeming weird, and exhaustively analyzed our sparse interactions with her immediately afterward for clues that she liked us and that our three-way nonrelationship was incrementally advancing. But we were definitely not gay! Nope—what?! Nope.

What would it have been like to be Sarah in high school? I marveled to myself on the drive home from our hangout. To be a constant object of interest and fascination in that way? As I visualized it, I felt an attack of envy, misplaced or not. Imagine feeling from a young age that you were that special, I sighed self-pityingly—that you could rely on people to notice and admire you, that you carried with you permanent resources to fell everyone in the room?

Then I stopped myself. Wait. *Imagine* feeling like that? Please.

I'd reached the lights; I leaned back in my seat a little and smirked. When I got the green and hit the pedal, a surge of confidence and self-satisfaction swept through my body, with the 1990s written all over it.

I *knew* what it felt like, or rather, *we* did. Julia and I had massive social power in high school—we were the twins.

Our glory reached its height at the end of our senior year, when our percentage grades in New Zealand's national exams reduced miraculously to the exact same aggregate number and we made a wide sweep between us of every humanities prize awarded by our school. At the end-of-year awards ceremony, we took turns walking back and forth across the stage to accept our haul of trophies, as the delighted student body increasingly lost its shit. When we advanced together at last, to accept our figurative crowns as joint "dux" of the school, there was so much stamping, clapping, and screaming that it seemed as if no one in the town hall, or maybe the entire city, was uninvolved. This for essentially a nerd prize.

New Zealand being a small country, our local fame quickly became national. Julia and I were plastered over newspapers, discussed on the radio, and featured on the six o'clock TV news. Our city paper reported, "Twins Julia and Helena de Bres, joint dux of Wellington East Girls College, have given up trying to be different from each other. It just doesn't pay off for the 17-year-olds, who are identical apart from the fact that one is left-handed and the other right-handed. 'We tend to be interested in the same things. If one of us tries to break away, it usually doesn't work out,' Julia said." ("Aghhh," Julia yelped when I located this and read it to her recently. "I can't believe I said that! Did they make me say that? Gross, gross, gross!")

That moment was the first time our celebrity status as twins moved beyond our immediate social circle to encompass the fascination of people we hadn't met. But, like all identical twins, Julia and I had been minor celebrities at the local level for as long as we could remember. Strangers stopped our parents in the street to coo over our stroller, we were all of our relations' favorites as kids, and we enjoyed a reliable supply of best friends, party invitations, and devoted teachers. When our cousin enlisted us as eight-year-old flower girls and outfitted us in matching apricot princess dresses, we coolly dominated her whole wedding. Everyone at our schools knew who we were (though not necessarily who we *each* were), and though we hung out with our fellow nerds, our twin status gave us a lifetime backstage pass to semicoolness. When I return to Wellington now, two and a half decades after graduating from East, I still have random middle-aged women rush up to me and gasp, "Excuse me—are you one of the de Bres twins?" The last time this happened, I was butt naked in a swimming pool locker room, and the inquirer turned out to be our high school principal's ex-secretary. "Why, yes, I am," I replied, slowly raising my bargain-basement towel, with the majestic self-possession of a starlet accosted by the paparazzi.

. .

The social power of twins derives from what the therapist Joan Friedman calls "the twin mystique": the awe and wonder our simple doubled-up-ness evokes. That mystique is grounded in a set of interests and needs that singletons think or feel twins can satisfy. Our power, like the power of women like Sarah, lies in our status as objects of desire, though what exactly people want from us isn't so clear. What do singletons think twins are for?

Some answers to that question I've already covered in this book. Twins have been used in many times and places for intellectual purposes, as tools to think with. In art and myth, they're employed as symbols of duality, wholeness, trickiness, creativity, social conflict, and perfect or pathological love, and as ways to explore distinctions between self and other, mind and body, male and female, and similarity and difference, period. In science, twins are central to behavioral genetics and clinical medical trials. At the more personal level, individual singletons enlist twins in their ongoing task of self-definition, using them as contrasting models when deciding how to act and be.

Singletons also use twins, or at least the idea of twins, for more emotional purposes. The early psychoanalyst Dorothy Burlingham diagnosed the common childhood fantasy of having a twin as a response to the trauma of maternal separation. The singleton toddler, she argued, misses the sense of symbiosis they had with their mother and conjures up an imaginary second self to merge with in her place. Burlingham also saw the twin daydreams of singleton children as externalizing a sense of inner insufficiency, conflict, or ambivalence. The fantasy twin can be everything the child dreamer wishes they were: the self who's perfect and blameless, or the one who isn't but gets away with it. Finally, the fantasy twin functions to imaginatively magnify the child's powers and protect them from danger. In Burlingham's words, the singleton kid thinks, "I am small and weak in the face of dangers, but if I were twice as big, twice as strong, there is nothing that I would not be able to do."

Variations on these emotional uses of twins persist into adult fantasies, too. The ancient Romans believed that each singleton is accompanied through life with a spiritual protector, or "genius," essentially a supernatural twin. When Duke Solinus comes across the two pairs of twins in Shakespeare's *The Comedy of Errors*, the duke marvels, "One of these men is *genius* to the other; and so of these. Which is the natural man, and which the spirit? Who deciphers them?" This fantasy has its dark counterpart in Jung's idea that singletons each have a "shadow self," made up of the parts of their character they'd rather not acknowledge, an idea the nineteenth-century doppelgänger literature took to its Gothic limit.

Twins are also objects of desire within the entertainment industry and the consumer market more broadly. They're regular fixtures in books and movies and on TV and social media, and they're popular advertising tools, given their rarity, cuteness, and quasi-scientific potential. They feature regularly in advertisements for twin packs, two-for-one deals, twin sets, twin engines, and twin appliances, and as co-twin controls for any product. (Nina used the cream and Dina didn't—look at Nina now!)

When I see twins being used in these ways, I often feel uncomfortable. I get why singletons do it, and their doing it has frequently been to my benefit. It's nice to feel special, to know you have something other people want. But, to revert to proto-feminist tween me, it's perilous to pin your value on the functions you perform for others. You're wanted, but not often for your full range, and any power you derive from the transaction is dependent on the goals of the desirer, not you.

In the case of women, this often gets framed in terms of objectification. One way to objectify someone is to treat them instrumentally, as a tool for your own purposes. Another is to deny their autonomy by acting as if they have no significant aims of their own. Beyond those core moves, Martha Nussbaum distinguishes a menu of distinct but often correlated options. The objectifier can treat the target as passively inert, interchangeable with others, violable, ownable, or lacking in subjectivity (with no thoughts or feelings worthy of note). Rae Langton adds three more moves: the reduction of a person to their body, the reduction of a person to their appearance, and silencing (acting as if someone is incapable of speech). All these acts fail to treat the target as a fully-fledged human and treat them instead as a thing.

If this analysis is right, twins have been regular targets of objectification since, well, forever. Twin objectification is most obvious and intense when it overlaps with objectification of the more familiar—and much more troubling—kinds. (Julia wants me to call this twintersectionality, but I'm resisting!) Take the common male sexual fantasy of sleeping with a pair of female identical twins. Those who are turned on by this scenario struggle to account for its seemingly evergreen appeal. One interviewee in *Salon* "clarified," "There are two of them and they are the same and I can have sex with them at the same time and did I mention that they're the same?" But the explanation is surely the dynamite pairing of the objectification of women, which many find arousing, with the objectification of twins, also hot. A straight male contemplating the standard singleton threesome can *sort of* treat his two partners as fungible—they're women, how different can they really be? But how much easier to objectify women if they're literally interchangeable, like, you know, twins are?

Wrigley's drew on this trope for decades in the series of female twins they employed to advertise Doublemint chewing gum. The Doublemint twins began in 1939 as stylized illustrations, but transformed into increasingly sexy flesh-and-blood models as the sixties and seventies went on. Somewhere in there, Wrigley's advertisers added the slogan "Double your pleasure," in case consumers had somehow missed their point. The twins in a threesome are there for the guy's gratification, not their own. If the fantasy includes the idea that the twins are also hot for each other, that's just part of the fetish. Though real-life twins are almost never sexually attracted to each other, the thoughts and feelings of the twins in the fantasy don't matter.

A much darker meeting of twin objectification with other objectification occurred in Mengele's experiments on twins at Auschwitz. Between 1943 and 1944, nearly fifteen hundred sets of twins imprisoned at the camp were subjected to a series of horrific violations, as part of Mengele's unhinged program of genetic research. The doctor injected dyes into twin's eyes to attempt to change their color, sewed twins together to create conjoined pairs, castrated or sterilized twins to try to change their sex, and planned to have male twins impregnate female twins to see if their children would be twins, too. He often used one twin in a pair as a control for the other, exposing the primary victim to contagious diseases, poisons, or surgery without anesthesia. He personally killed his subjects on operating tables and dissected them to compare their organs. Out of the original three thousand individual twins, less than two hundred survived. As a "scientist," Mengele was already inclined to objectify twins as resources for his "research." His companion inclination, as an antisemite, to objectify Jews as fungible, ownable, and violable led him to abandon all moral constraints in doing it.

A similar dynamic was at work in a less appalling form in the employment of twins in the eighteenth- and nineteenth-century freak show. Across Britain, Europe, and America, conjoined twins were exhibited at fairs, circuses, theaters, and museums as "living curiosities," alongside mutant animals, fat women, non-Westerners, intersex people, and those with unusual heights, prominent breasts and asses, or simply a lot of hair. Pretty much every mode of objectification was on display on the Odditorium circuit. Many freak show twins were treated as fully or partly the property of their "handlers," obliged to perform on demand for only a fraction of the proceeds generated by their show. Their audiences failed to respect their physical boundaries, poking, pinching, or even punching them. The twins' costumes and acts were designed to accentuate their similarity, and playbills reduced their humanity to the single tag of their twinhood. Doctors and professors were enlisted to introduce and pronounce upon them, as if they were zoological specimens, as if their own voices didn't suffice. And everyone exhibited a compulsive preoccupation with their body parts, including those ordinarily out of the decent conversation range for a Victorian. Here it was the intersection of conjoined twinhood and disability that greased the wheels of dehumanization.

The extreme forms of twin objectification tend to piggyback on these other, more familiar kinds. But twins don't need to have an independent marginalized social identity to be objectified in more minor or subtle ways. Whatever our gender, ethnicity, or body type, twins who look similar are forever objects of the singleton gaze. In daily life, strangers reduce us to our appearances, gloss over our subjectivity, and view us as interchangeable. (A mother told me that when one of her small twins fell over, a school friend looked down at the twin crumpled on the ground and said, "Where's the other one?") In Hollywood, most twin plots and shots focus obsessively on our physical similarity and our supposed swappability, as in the classic midcentury switcheroo films I discussed in chapter 1. In ads, too, we're usually identically dressed and posed side by side, as if the copy on the right has simply been cut and pasted in from the left.

Then there are the many ways I've mentioned that singletons don't just gawk at, but use us. Mythological and artistic representations of twins may not portray us negatively, but they often bleach out or distort our personalities and relationships to fit them for singleton purposes.

Likewise, the overwhelming majority of scientific twin studies don't explore twinship itself, the relational aspects of twinhood that matter directly to us twins. Instead, we—or rather, our genes or highly specific aspects of our bodies and behavior—are commandeered for research goals that have nothing to do with twins per se. Films rarely provide a realistic picture of what twinship feels like from the inside either; instead, we're displayed for the voyeurism, laughs, or pleasurable horror of non-twins. In ads, often the only visible difference between us has been produced by an external feature, the product we're hawking: our supposed identicalness serves as a blank slate for capitalism to draw its face on.

• •

I was thinking about all this on a visit to Julia last year, while sitting in the sunny window nook in her kitchen. I'd spent a week reading about circus sideshows, the male gaze, feminist approaches to porn, the My Twinn dolls of the nineties, the female nude, et cetera, stewing in all the parallels between the objectification of women and the objectification of twins. I was crouching over my laptop in the summer sun, squinting at the latest item on my reading list, and preparing to get het up about it, when Julia's daughter, Flo, sauntered into the room.

"What are you reading about, Lena?" she asked.

I explained, warily. Flo, like me and unlike Julia, is a natural philosopher. She nerds out in the philosophy electives her middle school offers and quotes the Stoics to me while I'm cooking or we're swimming. I prefer not to talk about philosophy with my family because I see it as work and them as not. Flo knows this, so she likes to trap me in a conceptual debate that I'll get super invested in and be unable to drop. She'll have a blast, partly because she enjoys philosophizing, partly because she's defeated me yet again by making me do it, and I'll come out feeling exploited, exhausted, and crushed. It's one of our dynamics, and part of me loves it.

"But, Lena," Flo said, extending her body along the adjacent window seat, like an ancient Greek settling in for a symposium. "We *are* objects."

"Sure," I said reluctantly, "okay, yes."

"So what's your problem?" Flo raised a finger. "What's wrong with treating someone *as the thing they are*?"

Part of what puts me on the back foot in these interactions, alongside Flo's preternatural intelligence, is that she always looks incredible. Her fashion sense is Gen-Z-anime-disco-dramatic. I saw I was about to be backed into a corner by an eleven-year-old gamer in purple glasses, a pink hoodie with bunny ears, and leopard-print tights.

"I'm not doing this, Flo!" I wailed.

"Yes, you are, Lena," she replied with satisfaction. "You're doing it."

Soon enough I was, and to Flo's additional pleasure, I had to admit she had a point. All humans are objects, in the sense of "physical bodies extended in space and time and subject to the laws of nature," as well as in the sense of "targets of the regard and actions of others." Each of us is a material and a social being, and it'd be weird to morally object to someone else recognizing and responding to those essential facts about us.

Flo's defense of objectification becomes even more compelling when you note that some instances of treating people as objects in each of the ten ways I listed earlier are ethically neutral. Nussbaum gives the example of using your lover's stomach as a pillow. You're instrumentalizing them, sure, and treating them as an inert, fungible body into the bargain. But, as long as they have no objection to your head being where it is, you're not at fault for putting it there.

Nussbaum takes this point further by arguing that some objectifying acts aren't just morally neutral, but positively valuable. Her central example is the kind of objectification that regularly happens in sex. Sexual passion tends to obsessively focus on the other person's body and disregard their other aspects, such as, say, their capacity to understand Kant's Second Critique or advance the climate revolution. Sex also involves loss of control, a temporary disregard for autonomy and subjectivity, and the breaching of significant physical and emotional boundaries. Those are all marks of objectification, but they're surely valuable aspects of sex—not just benign features of it, but part of what makes it great.

To take this defense of objectification to the end of the line, some have argued that *not* treating someone as an object can be a moral failing. One underacknowledged aspect of social marginalization is not being physically objectified enough. The patriarchy hypersexualizes women, sure, but it also undersexualizes them, discouraging them from displaying or welcoming sexual desire and treating their bodies as embarrassing or disgusting—especially if they're over fifty or fat. Straight culture can make it hard for gay, lesbian, trans, and intersex people to feel physically attractive. When they're not made invisible, their sexuality is stereotyped as effete (gay men), incomplete (lesbians), or threatening (gender-diverse people). Racist Western culture does something similar to Asian men, casting them as sexless and undesirable. Leslie Green argues that the injustice suffered here is actually subjectification, not objectification: having your nature as a full-fledged embodied human systematically ignored or denied.

Something similar can be said for the instrumentality mode of objectification. Elderly and disabled people aren't just treated as sexless, but as disposable across the board. Because most of us badly want to feel useful in some way and ground our self-respect and identity in our capacity to cooperate with others, this experience is often very painful. The widespread failure to instrumentalize the retired, unemployed, and disabled isn't a way of respecting their humanity, but a potent form of dehumanization.

Flo's challenge sent me back to Nussbaum's take on the ethics of objectification, which was a good thing because it's helped me get clearer on my long-standing queasiness about singleton uses and portrayals of twins. *Thanks, Flo.*

Nussbaum's key suggestion is that the deciding factor in assessing objectification is its compatibility with mutual respect and equality. Some acts—hypersexist art and advertising, say—will flunk this test, but others won't. To work out which, we have to consider both the broader relationship the act occurs in and the meaning it holds for those involved. Objectification might be *temporary*: a momentary part of a relationship that's respectful overall. Or it might be *symmetrical*: both people objectify each other, so the individual acts don't reveal an unequal power structure. Or the circumstances surrounding an act could indicate it's *consensual*: genuinely wanted by all involved, so not a dismissal of the target's inner life. It'll depend on the details, but in many of those cases, the objectification won't be morally troubling because the person isn't being seen or treated *only* as an object: their full humanity is being acknowledged and respected at the same time.

If we take this line, the ethics of twin objectification is a mixed bag. Some forms of it completely violate the fundamental respect and concern we owe all humans. Mengele's program is a case in point, as is the traditional use of twins as sacrificial animals to appease the gods in Guinea-Bissau, where, according to Alessandra Piontelli, "nobody [sees] any differences between twins and goats."

In contrast, the instrumentalization of twins in mainstream contemporary science seems acceptable. Genetic and clinical studies of twins have to pass ethical review, which minimizes risks of harm and involuntary participation. If the studies only focus on (parts of) twins' bodies, they do so for nonarbitrary, defensible reasons, and they build in a fundamental form of reciprocity, given that the results of twin research are just as likely to benefit twins as singletons. Most of these studies, while they have objectifying features, overall respect the humanity, autonomy, agency, nonfungibility, boundaries, inalienable rights, and subjectivity of their twin subjects.

The intellectual and aesthetic uses of twins in myth, art, and entertainment are ethically more complicated. In their defense, all myth and much art involves abstraction and exaggeration, to heighten its symbolic and formal qualities. So narrative simplifications of twinhood aren't necessarily gratuitous or intended or interpreted as representations of what twins are really like. And the work twins do in these genres—exploring questions about human identity, relationships, and experiences—is, like the work twins do in science, something twins can benefit from, reducing the sense of an unequal exchange.

There's a good argument, though, for supplementing the more abstract and simplistic cultural portrayals of twinship with more fleshed-out representations. While there's nothing troubling in principle about portraits of female nudes, it *is* troubling when the overwhelming majority of female nudes are presented in objectifying ways (as they are). The same could be said for the predominance of washed-out and distorting cultural images of twins. The International Council of Multiple Birth Organisations lists the depiction of twins as "depersonalized stereotypes" as one of the wrongs in its "Declaration of Rights and Statement of Needs of Twins and Higher Order Multiples," and that strikes me as right. I'm not going to claim that twin jokes, twin threesome fantasies, and a lifetime of singleton side-eye are the great social justice issues of our time. But depersonalization remains a risky road for any of us to travel down, even when we're walking it two by two.

• •

Was it the prospect of being treated only as an object that bothered me as a sex-negative teen? Maybe, but if I'm honest with myself, I think something else was going on, too. I suspect that junior me didn't want to be treated like an object *at all*, even in the allegedly good ways. I didn't want to be reminded that I had a body: I wanted to lead always with my strong suit, my head. I have the feeling that for me, getting morally worked up about objectification has sometimes been a cover for that anxiety about my body, a political sublimation of a suspect metaphysics of personhood.

The philosopher Ann Cahill doesn't like Nussbaum's account of the ethics of objectification for this reason. The idea that any treatment of someone as an object has to be hedged with reassurances that they're not *just* an object gives the impression that the physical and social dimensions of ourselves—our status as material things, our perpetual vulnerability to the regard of others—are inherently problematic. We might have to acknowledge those dimensions, Nussbaum's view seems to say, but we need to carefully manage them, lest they take over, in our minds and those of others, what really matters about us: our intellectual and emotional capacities, rather than the physical properties we share with nonhuman animals.

Cahill rejects this Kantian account of what a person is, insisting that our bodies are no less central to our selves, our worth, and our dignity than our minds are. She argues that Nussbaum's assumption that objectification always requires justification implicitly denies that, thereby vilifying our bodies. But the problem isn't just Nussbaum's account, Cahill suggests: it's the metaphysical baggage of the concept of objectification, period. Intentionally or not, all those who appeal to it smuggle in a picture of the body as inherently degrading, a picture we'd be better off without.

Cahill doubts the usefulness of the idea of objectification for another reason, too. She suggests that most alleged examples of it don't really treat people exclusively as objects. Much sexist pornography, for instance, doesn't portray women as passive, inactive, and mindless. The women clearly have thoughts, feelings, and desires, and part of the kick viewers get is precisely that of seeing not a mere thing, but a live human with her own agency and will being treated as if she were worth nothing. The problem here, Cahill argues, isn't objectification but what she calls *derivativization*. (An ugly word for an ugly thing.) Women in sexist porn are treated as important only to the extent that they reflect and satisfy the desires of men. They can be subjects, yes, but only if they truncate their subjectivity so that it perfectly meshes with their proper function as male gratifier. Their only significant desire is a desire to please men, and their whole reason for existence is reduced to that one subservient role. GIRLS CAN DO ANYTHING, as a peppy poster in my 1980s bedroom said—provided it's something men want.

• •

Whatever we might say about the ethics of objectification and derivativization, people do objectify and derivativize women, and singletons do objectify and derivativize twins. How, then, should we at the receiving end respond? One popular option is to make use of our situation, to instrumentalize our instrumentalization and, indirectly, those who instrumentalize us.

When Julia and I jointly won the academic medal for the school of Humanities and Social Sciences at the end of our college degrees, we were invited to give the single student address at our graduation ceremony. I'm not sure what the organizers had in mind, but we decided to lean into the opportunity and deliver our speech in alternating sentences. We threaded in twin references from every branch of human achievement, made faux-hostile digs at each other, and suggested that our entire capped-and-gowned cohort looked like the outcome of one freakish multiple birth.

It was gratifying to witness that crowd of enthralled singletons hanging off our every word, laughing at our stupid jokes, and high-fiving us as we retired with faux bashfulness from the stage. It felt like the triumphant culmination of the countless times we'd hammed up our twinhood over our twenty-one years: the time we jointly submitted our handwriting to be analyzed by a magazine graphologist, the time we auditioned to be twin presenters of a teenage breakfast program, the time we were extras for a TV trivia show featuring teams of twin contestants. And all the smaller performances of similarity, opposite-ness, "twintuition," and perfect companionship that we'd rolled out for our social profit to friends, family, and strangers, since around the age of three.

The performance of twinhood is part of twinhood, though I'm thinking not all twins are this much of a dork about it. Still, to stick with our fellow dorks for the moment, there's that famous annual twin festival in Twinsburg, Ohio, where every August hundreds of sets of twins turn up to participate in costumed look-alike contests and parade through the center of town. While the Twinsburg twins seem to be mainly in it for a good time, other twin performers are more clearly out for fame and fortune. The VH1 reality show *Twinning* featured twelve pairs of identicals competing to communicate telepathically, stay the longest inside tilting makeshift "wombs," and live in separate houses for weeks without falling apart, for a series-end "twinners" prize of $222,222.22. In the 1990s, Debbie and Lisa Ganz ran a lucrative Twins Restaurant in Manhattan, decorated with double barstools, light fixtures, and doorknobs, and staffed exclusively by identical twin waitstaff whose contracts only allowed them to work if both twins of a pair were on shift.

Some female twins crank up the threesome fantasy as part of their shtick. In their "Boy Toys" project, the L.A. "artist celebrities" Lexie and Allie Kaplan posted photos of themselves on Instagram posing seminaked in bed with plush action figures they successfully sold online for $333 each. The Olsen twins ended their co-host stint on *Saturday Night Live* with Mary-Kate yelling, "Remember, we're legal in three weeks!" Other family-friendly female twins pimp themselves out more indirectly. The Texan twins Brooklyn and Bailey McKnight have had a YouTube channel since the age of thirteen: a spin-off from their starring role, from age nine, on their mother's channel, Cute Girls Hairstyles. After being, one might argue, objectified by their mother, they turned enthusiastically to objectifying themselves. Their content—wholesome, upbeat videos about their everyday lives going to school, applying makeup, buying clothes, dating boys—has garnered them five million subscribers, who they've cannily converted into consumers via a line of merchandise, partnerships with JCPenney and Baylor University, and ads and endorsements. "I guess just the goal in general," Bailey says, "is to continue to expand and grow upon the brand and the business that we've already built, and the empire that we have created for ourselves, basically."

The frequency with which twins perform their twinship for personal gain should give us pause when we're tempted to construe the conjoined twins of the nineteenth-century freak show as passive tools of their showmen employers. Chang and Eng Bunker were initially highly controlled by the British businessman and American captain who bought them from their parents, after coming across them in a Siamese fishing village in 1811. But they quickly proved to be savvy businessmen themselves and ended up operating their own tour, making enough money to retire as comfortable, landowning gentleman farmers in North Carolina. Robert Bogdan has argued that many freaks successfully played the game this way, adroitly turning their audience's fascination with them to a tidy profit. The point extends to the reality-show, restauranteur, and influencer twins of our day, and, in less flashy ways, to twins just going about their ordinary lives. All twins know how to perform what Karen Dillon calls "the spectacle of twinship" when it's to their advantage, taking up the various social caricatures of their relationship and stoking up the cultural fantasies associated with them. When dealing with singleton fascination, we tend to know exactly what we're doing.

What *are* we doing, though, at a more fundamental level? Making the best of a bad situation, or transforming a potentially bad situation into a perfectly good one? That question comes up quickly in relation to women in the class on pornography I teach in my introductory ethics course. My students all identify as women, trans men, nonbinary, or genderqueer, and attend a progressive liberal arts college in the northeastern United States. They seem to have been ejected from the womb as third- and fourth-wave feminists, and, in the early years of my job, getting them to see why a feminist might be even slightly worried about pornography used to take so much mental and emotional energy that I'd have to gear up for it by promising myself a strong drink later in the evening.

The second-wave reaction to women pandering to the male gaze was to, at best, excuse it. A major question of Simone de Beauvoir's *The Second Sex* was why women seemed to participate so actively and enthusiastically in their own objectification. Her answer pointed to the difficulty of their renouncing the few advantages patriarchal oppression affords them. Some of these advantages are material: the financial support and physical protection that cooperation with men supplies (admittedly, support and protection that women largely need due to the prejudice and aggressions of other men). But another big advantage, according to Beauvoir, is existential. Women who acquiesce to their social role as passive object don't have to face the bracing metaphysical challenge of assuming full responsibility for their lives. Their role as semi-human may oppress them, but their dependence on men also spares them the terrifying burden of wrenching some value and meaning out of the cosmic void. Succumbing to the temptation to forgo authenticity and remain merely a thing is, for many women, overwhelmingly easier than fighting it.

Andrea Dworkin later argued that internalizing the male gaze also has the benefit of sparing women from the horror of staring misogyny straight in the face. It's not fun to recognize that many of your most important relationships may be infected by fundamental disrespect and backed up by the threat of force. Sandra Bartky added that, as well as preventing some serious losses and costs, self-objectification has a positively enjoyable side for at least some women. Those who successfully live up to the exacting norms of ideal female beauty and femininity can derive a strong and satisfying sense of pride and mastery from doing it. They can also find it hot, given the way that (as Catharine MacKinnon argued) the patriarchy eroticizes female objectification.

All of this requires some false consciousness on the part of women, these second wavers tell us. At some level we women know that we're existentialist subjects as well as objects; we know that our desires are corrupted by the patriarchy; we know that the cards, however appealing in the moment, are ultimately stacked against us. Lying to yourself isn't great, nor is being trapped long-term in this sexist dynamic, which has terrible consequences for everyone. But it's easy to see why women might make their compromise with the system this way, wittingly or not. The cultural pressure is immense: "Women live in objectification," MacKinnon wrote, "like fish live in water." The prospects of a single individual shifting the entire system are nil, while the prospects of that individual making her life significantly worse in the attempt to do so are obvious. Given all these facts, while we feminists might find female self-objectification unfortunate, we might also consider it excusable. It seems mean-spirited at best, and victim blaming at worst, to criticize women for taking what little power they can muster in a social world largely stacked against their freedom and equality. Give the poor things (sorry, "subject-object hybrids") a break!

This makes a lot of sense to me. When I started teaching in the late 2000s, however, I found that many of my students considered it deeply and offensively misguided. What they lived in like fish live in water was the idea that what women do with their bodies and relationships is no one's business but their own. The idea that women who enjoy self-objectification are mistaken about their own best interests or self-deceiving sounded patronizing, even anti-feminist, to them—just another instance of women being told that others know better than they do how they should think, feel, and act. The second-wave analysis also came across as sex negative, with its finger-wagging critique of what many of my students took to be the joyous and healthy expression of robust, unconstrained sexuality. What may look degrading to *you*, they wanted to say, isn't degrading at all, once you understand it as the outcome of the authentic exercise of a woman's autonomy. We women aren't mere dupes of the patriarchy, they insisted; we make active use of its conventions, for our own pleasure and material advantage. As part of this, some women may adopt the appearance of submission playfully or transgressively, not to conform to or endorse, but to subvert the norms that once oppressed them. In this way, Nadine Strossen writes, "Words or images that literally depict a woman's powerlessness may well have an empowering effect on female viewers."

A similar line of argument appears in the branch of disability theory that Michael Chemers christened "freak studies." In his manifesto for the field, Chemers argued against the traditional understanding of freak show performers (including conjoined twins) as exploited and oppressed victims, condemned to helplessly reflect and reinforce the prejudices of their audience. Instead, he insisted on their status as performers: actors who savvily constructed an exaggerated display of abnormality as a lucrative form of entertainment. "Freakery," Chemers argued, was a "highly specialized and potentially liberatory form of performance art." Its effect wasn't simply to reproduce dominant social discourses of normality and deviance. Instead, in displaying their agency, and positioning themselves as both wonders of nature and successful entrepreneurs, freak show performers at least partly subverted their audience's received views. "The exploitation of the freak show was perpetrated not by the audience, but by freaks themselves." As a result, those who advocate shutting down freak show performances to protect freaks from objectification risk putting in place "new restrictions that strive to contain the very body they have ostensibly been created to liberate."

This cross-disciplinary brand of empowerment theory can take one of two lines when it comes to self-objectification. It can argue that certain acts of apparently extreme self-objectification—say, acting as the victim in a pornographic rape scene—are indeed objectifying, but none the worse for that, provided that those involved perform them enthusiastically. (This might be classed as an example of the temporary and consensual kind of objectification that Nussbaum claims is ethically unproblematic.) Or an empowerment theorist can argue that the concept of self-objectification doesn't even make sense. Whenever you attempt to objectify yourself, this idea goes, you necessarily fail. In actively trying to make yourself into an object, you can't help but reveal yourself as an agent—a subject, not a thing.

I used to find this take so alienating when it came up in my ethics class that I didn't know what to do with my face while standing by the chalkboard listening to it. Plenty of things my students say I disagree with, while still being able to imaginatively project myself into their point of view. This was different: I just could not imagine how they could really believe what they were saying. But it was worse than that. You might think that my inability to understand where they were coming from would have reinforced my conviction that my take was right. But it didn't! I didn't feel at all confident that my resistance to my students' position was well grounded, because what my mind was mainly attending to throughout the discussion was the apparently giant gulf between their experience of sex and my own.

The way they were talking about it, they genuinely saw their sexuality as a source of power: they witnessed its effects in the world, leaned in more closely to examine the phenomenon, taught themselves how to manipulate it for their purposes, then did so and won. I had never viewed my sexuality in that way. I saw it as a *site* of power, sure, but someone else's, not my own. My students reminded me of Clare in the basement, enthusiastically poring over her parents' erotic photo shoot as if it had something useful to tell her, or Eliza liberating herself from her mother by liberating Ken from his denim cutoffs. They were gleefully pole dancing like Katie, slaying men in the world's major metropolitan areas like Sarah. By comparison, they made me feel like a kid again: that anxious girl out in the suburbs, up on the windy cliff, frightened and ashamed of her body, unsure she'd ever really earn her assigned gender. I couldn't in good conscience tell myself, like a good second waver, that my students' power was an illusion. They did seem powerful. I could tell they did because of how powerless listening to them made me feel.

• •

A few things have changed about this awkward teaching scenario over the past few years. One is that I've been thinking more about twins. I still don't relate to the sense of empowerment that some of my students get from sexually objectifying themselves. I haven't done a lot of that myself, though I've pervasively derivativized myself in my relationships with men. But when I think of the instances when I objectified myself as a twin teenager, I do sort of get it. While I never knew how to put my sexuality to social use, my twinhood was an ever-ready arrow in my bow. When I used it, when my plays for singleton attention smoothly and predictably hit their targets, it felt so good.

Another pair of changes spring from the general raising of feminist consciousness we've experienced recently, in the wake of the #MeToo movement and in response to the unveiled sexism that flared up everywhere during the Trump administration. Like many of us, I've spent more time over the past few years than ever before thinking about the roles that sex, gender, and the patriarchy have played in my life and the lives of others. I see clearly now how much of my sense of powerlessness about my body springs from sexist and ableist views that I fully reject when I actually unearth them. Against this backdrop, I find my students' persistent focus on agency and empowerment hopeful. Even when I have my doubts about it, I'm happy to hear it. We could all use a little empowerment over here.

At the same time, I've noticed what looks like a shift in my students' views, bringing them closer to mine as I inch closer to theirs. The core idea behind empowerment feminism and the related discourse in disability studies is that we should accept all desires as given, and morally evaluate behavior solely on the basis of consent. You (or your audience) want what you want, however weird or disturbing it may appear to others, and provided that everyone fulfilling those wants is into it, there's no valid space for ethical critique.

This emphasis on consent is important: no one in this debate doubts that consent is morally necessary for most interactions and exchanges, including all of those in the sexual realm. But the disagreement is over whether it's sufficient, and here's where feminists and disability theorists of the older school will balk. We can surely sometimes consent to doing things that aren't self-respecting or don't promote our well-being, simply because they're the best out of a range of bad options. Similarly, we can genuinely desire something while under different, and more just, social conditions we would have desired quite other things. Taking these facts seriously, as Amia Srinivasan has recently forcefully argued, means opening desire up for political critique.

My sense is that my students are much more open to this idea than they used to be, which I'm inclined to chalk up to the recent resurgence of public debates about feminism. They're still understandably resistant to criticizing women's erotic desires and choices, given the long history of literally everyone doing that. But they seem more receptive to the thought that feminism kneecaps itself if it doesn't scrutinize the content of female (and male) desire. We know that our beliefs and preferences are formed within a system that, despite all the feminist advances of the past half century, remains deeply sexist. Wouldn't it be nothing short of a miracle if they came out of that dirty wash clean?

For that reason I continue to think that the concept of self-objectification makes sense, and that many of the examples of it prevalent in our sexist culture are morally troubling. For the same reason, though at a much-lower pitch, I can't feel totally comfortable with the self-objectification of my fellow twins and me. A lot of the hamming up of twinhood that twins do in everyday life falls into the benign or positive category. It can poke fun at, and satirically subvert, the objectifying clichés singletons harbor about twins. Performing those clichés, even unironically, can provide a genuine sense of power and status for a group of people who are often made to feel freakish. But when twins ratchet up the objectifying qualities of their act and take it public, my ambivalence increases. Stoking the fire of the singleton gaze is a low-level political act, with consequences for twins everywhere. I can see why twins might want to do it, but I resent it, too. I'd prefer to see my fellow twins and me rejecting this persistent othering of ourselves, and of other people. To mangle a quote, none of us are free from othering till everyone is.

The idea that membership in an objectified group will make you more respectful of the full humanity of other objectified people is often overoptimistic. One of the more depressing facts about Chang and Eng Bunker is their enthusiastic ownership of twenty-six slaves in the pre–Civil War South. René Zazzo reports a similarly depressing instance of a soldier getting his twin brother to impersonate the soldier with his girlfriend during his monthslong absence, to keep her faithful to "him." (Zazzo calls this "touching"; I'd call it rape.) These examples of twins instrumentalizing themselves only to assert dehumanizing ownership over others are a reminder that interminority solidarity can be elusive. But we twins and singletons could continue to hope and work for it together, couldn't we?

. .

When I was last in New Zealand, I got a surprise Facebook message from my high school prom date. "Hi Helena," it said, "I have been cleaning out Mum's house and I came across this wee blast from the past. If you want it I could post it to you if you send your address."

Attached was a photo of one of the newspaper articles announcing Julia's and my nerd triumph at Wellington East back in 1996. When I suggested we meet for lunch, Alex explained that he'd saved the article when it came out "and good old mum never threw anything out."

I thought about this as I drove to the café a couple of weeks later for our meeting, the first since the prom two and a half decades ago. What had or hadn't been going on between him and me back then? I asked myself. Alex was a couple of years older than me and had played the violin at the church my mother went to. He was half Italian, half Irish, with soft brown curls and big brown eyes, and I'd nursed a mini-crush on him from a distance for years. When the time came for the prom and I predictably had no date, he was the first person it occurred to me to ask. To my surprise, I managed to do it, and also to my surprise, he said yes. In the photos of him that night, wearing a tux and boutonniere in the family backyard, he looks great—and, more important, *so do I*. I'm a seventeen-year-old brunette in a sleek black shoulderless dress and I look gorgeous.

I'd already made my big move by asking Alex out and was too timid that night to make another. Alex was gentle and shy, too, and we hardly knew each other. Nothing happened. But it seems likely to me now that it could have if I'd been more confident in my nonexistent game. *Jeez, Lena*, I want to say to myself, looking at that backyard picture, *it wasn't that difficult! You could have nailed him!*

I didn't, but I did manage to get a photo of myself with my identical twin stored in his mom's basement for twenty-five years. My sex game may have been nonexistent in high school, but my twin game was killer. Leaving the café, I peered at the picture in the newspaper clipping: Julia and I are high-fiving awkwardly on our local beach, in our dork clothes. *High-five, Ju*, I said to myself as I got into my car. One of the nice things about twin power is that every social victory it secures is shared.

CODA

Roland Barthes writes in *Mythologies*, "The cultural work done in the past by gods and epic sagas is done now by laundry-detergent commercials and comic-strip characters." Twins, with their shape-shifting talents, have occupied both of these roles in human history. In the ancient Indo-European world, they appeared as the multicultural Divine Twins, astral-equestrian aiders of humanity and saviors of shipwrecked sailors. In the early twentieth century, they starred as one of the best-known trademarks in American advertising: Goldie and Dustie, the mascots of Gold Dust washing powder. LET THE TWINS DO YOUR WORK, the Gold Dust slogan read, accompanied by a picture in which one twin washed and the other dried.

What cultural work can twins do for us now? One idea running through this book has been that philosophizing about twins can help unseat a crusty and constraining model of what it is to be a person. In the West, many of us assume that we're each physically and mentally discrete: tucked up in our bodies and minds solo, wrapped in a single continuum of skin. We're attached to that idea because we ground our autonomy in it, along with our dignity and value. From this perspective, other people, however much we love them, are a major metaphysical and ethical hazard. Get too close, and they risk dissolving the bodily, mental, and emotional boundaries that make each of us who we are.

Twins put pressure on this picture of personhood in multiple ways. Physically, cognitively, emotionally, agentially, they tend not to see or treat each other (to quote the writer Helen Garner on humans more generally) as "discrete bubbles floating past each other and sometimes colliding." Instead they "overlap, seep into each other's lives, penetrate the fabric of each other." They offer a concrete, real-life example of how someone can be their own distinctive person while embracing the lifelong role of another in shaping their identity, values, and agency in the world.

Twins also show how appealing this alternative conception of personhood is. The traditional Western picture is surely right about something: intimate relationships can result in domination, even annihilation. But our culture tends to take that concern to extremes. Why think that autonomy and sociality are always opposed, that mutual enmeshment must ever reduce the self rather than expand it? Maybe if we were less defensive about our boundaries, less inclined to fear those who challenge our sense of who we are, we'd be less likely to marginalize and stigmatize those we see as different from us. And maybe if we saw ourselves as more capacious, fluid, and relational than we tend to, we'd be more likely to approach the full range of possibilities life offers us with curiosity, imagination, and courage. Twins—or the idea of them—can illuminate these possibilities and give both singletons and twins models of how to be more twin-like in their self-conceptions and relationships. We each contain multitudes, we're often told. What if we lived, really lived, like that were true?

AUTHOR'S NOTE

This is a work of creative nonfiction. I've changed some names, for anonymity's sake. Dialogue is reconstructed, but reflects my honest recollection of the content of what was said. Sometimes I've added some fictional details to flesh out a scene. Here and there, in ways I hope are obvious, I've exaggerated for comic effect. Otherwise the recounting of events and experiences is accurate, to the best of my knowledge.

ACKNOWLEDGMENTS

Biggest thanks to Julia—for everything, but also for collaborating with me on this book in particular. Julia was excited about the project from the beginning and was my first adviser, reader, and supporter all the way through. My discussions with her (and her mega-long voice memos) informed the arguments and improved the writing, and her fabulous illustrations have made the final product a true reflection of our lifelong partnership. We started working on the book at the beginning of the pandemic, when I was in California and Julia was in New Zealand. One reason I wanted to write it was that it would allow me to think about and talk with my twin regularly, when it seemed like everything else was falling apart. I can't say how much it has mattered to me over the past three years to have had something to attend to that represents a love letter to everything I find most precious and hopeful in the world.

Thanks, secondly, to my mother and father, Angela and Joris de Bres, for being such great parents of twins, and for letting me reinhabit my childhood bedroom for a year in 2020–21 while I wrote the first draft. This book has benefited in major ways from their recollections, feedback, and encouragement.

I'm hugely grateful to my agent, Tisse Takagi, for believing in this book, and me, and for all her help in getting it on the road to publication, and to Peter Tallack for ably representing me later on. I'm not only grateful, but delighted, to have had Callie Garnett edit it. I'd say it was a dream come true to nab an identical twin as an editor, but the possibility never occurred to me. (We're 0.3 percent of the population!) It's been amazing to check my thoughts on twins against Callie's experience, as well as to benefit from her expert advice and exceptionally close attention to the manuscript. Thanks also to Jillian Ramirez, Barbara Darko, Rosie Mahorter, Lauren Moseley, and everyone else at Bloomsbury for their enthusiasm about the book and skillful work in making it a reality.

Thanks to the insightful and generous readers who gave me invaluable feedback on drafts of the chapters: Sasha Eskelund, Carol Hay, Sharon Horne, Ingrid Horrocks, Victor Kumar, Serena Parekh, Katia Vavova, Julie Walsh, and the enthusiastic group of philosopher-writers I spent a week with at Quimby Country Lodge in Vermont. Thanks to Jonny Thakkar at *The Point* for publishing "It's Not You, It's Me," the essay that set off my thinking on twins. Thanks to Wellesley College for keeping me fed and housed while writing, and to baristas at the Botanist, Maranui, Vic Books, Chocolate Fish (Wellington); Verve, Coffeebar, Café Borrone (Palo Alto / Menlo Park); and

Curio, Hi-Rise, Intelligentsia, and Forge (Cambridge / Somerville) for keeping me awake. For fun and consolation, thanks always to my best buds, Gabrielle Baker, Hannah Cook, Candice Delmas, Corinne Gartner, Adam Hosein, Carmen Saracho, Liz Young, and my beloved fur twins, sweet Jasmine and feisty Juno.

NOTES

TWINS IN WONDERLAND

2–3 no two distinct things can *exactly* resemble each other: The principle, known as Leibniz's law, or the identity of indiscernibles, appears in book 9 of Gottfried Wilhelm Leibniz's *Discourse on Metaphysics*.

3 "If it was so": Lewis Carroll, *Through the Looking-Glass*, in Lewis Carroll, *The Annotated Alice*, 150th Anniversary Deluxe Edition, ed. Martin Gardner (New York: W. W. Norton, 2015), 213.

4 wonder is also where Socrates claimed philosophy begins: Plato, *Theaetetus* 155D, in *Plato: Complete Works*, ed. John M. Cooper (Indianapolis: Hackett, 1997), 173.

4 "If you think we're wax-works": Carroll, *Through the Looking-Glass*, 212.

7 "ARE THEY TWO MEN?": Nadja Durbach, *Spectacle of Deformity: Freak Shows and Modern British Culture* (Oakland: University of California Press, 2009), 63.

7 "the mathematics of personhood": Allison Pingree, "'The Exceptions That Prove the Rule': Daisy and Violet Hilton, the 'New Woman,' and the Bonds of Marriage," in *Freakery: Cultural Spectacles of the Extraordinary Body*, ed. Rosemarie Garland-Thomson (New York: New York University Press, 1996), 173.

11 the twin rate has roughly doubled: Laura Spinney, "The Twin Boom," *Aeon*, August 18, 2016. Spinney notes that,

like the fraternal twinning rate, the identical twinning rate appears to have been *slightly* raised by ART, for mysterious reasons. Twin rates continue to vary significantly across the world: they're lowest in Asia and highest in Central Africa. Benin wins, at twenty-eight twins per thousand births.

12 they're at somewhat greater risk when young: Alessandra Piontelli, *Twins in the World: The Legends They Inspire and the Lives They Lead* (New York: Palgrave Macmillan, 2008), chap. 11, "Abuse and Neglect." The apparent cause is the significantly heightened physical, emotional, and financial pressure placed on parents by the arrival of twins.

12 "Congress of human oddities": Visible in a photograph in Rosemarie Garland-Thomson, *Extraordinary Bodies: Figuring Physical Disability in American Culture and Literature* (New York: Columbia University Press, 1997), 51.

13 the twin "neo-gothic": Alessandra Piontelli, *Twins: From Fetus to Child* (Abingdon, U.K.: Routledge, 2002), 3.

13 "Good people live honestly": Fyodor Dostoyevsky, *Notes from Underground and The Double*, trans. Jessie Coulson (London: Penguin, 1972), 222.

13 to detect the workings of thieves: Elizabeth Stewart, *Exploring Twins* (New York: Palgrave Macmillan, 2003), 18.

13 "Calm down, breath of the twins": Kate Bacon, *Twins in Society: Parents, Bodies, Space and Talk* (New York: Palgrave Macmillan, 2010), 5.

14 "May you become the mother of twins!": James Rendel Harris, *Boanerges* (Cambridge: Cambridge University Press, 1913), 315.

14 involuntary scientific experiments: The documentary *Three Identical Strangers* (directed by Tim Wardle, CNN Films, 2018) dramatizes the consequences of the disturbing adoption

study conducted by Peter Neubauer and Viola Bernard on both twins and triplets in 1960s New York.

14 appalling violations: An alarming catalog appears in Piontelli, *Twins in the World*.

15 Elizabeth Grosz: Elizabeth Grosz, "Intolerable Ambiguity: Freaks as/at the Limit," in Garland-Thomson, *Freakery*, 64.

17 Rosemarie Garland-Thomson: Garland-Thomson, *Extraordinary Bodies*, 65.

18 are almost genetically identical: Nancy L. Segal, *Twin Mythconceptions: False Beliefs, Fables and Facts About Twins* (London: Elsevier, 2017), 37–38, 69–71.

18 *second* and *other* are simply the same: Carl Jung, *Psychology and Religion: West and East*, vol. 11, trans. Gerhard Adler and F. C. Hull (Princeton, NJ: Princeton University Press, 2014), 118.

22 "But I don't want to go among mad people": Lewis Carroll, *Alice's Adventures in Wonderland*, in Carroll, *Annotated Alice*, 79.

23 "looked at her in a secret way" and "Do you think I will grow into a Freak?": Carson McCullers, *The Member of the Wedding*, in *Collected Stories of Carson McCullers* (Boston: Houghton Mifflin, 1998), 272–73. See the helpful discussion in Rachel Adams, *Sideshow USA* (Chicago: University of Chicago Press, 2001), chap. 4, "A Mixture of Delicious and Freak: The Queer Fiction of Carson McCullers."

23 "we are all androgynous": James Baldwin, "Freaks and the American Ideal of Manhood," *Playboy*, January 1985.

23 "the we of me": McCullers, *Member of the Wedding*, 378.

1. WHICH ONE ARE YOU?

32 Sexy/Frigid Twin divide: Juliana de Nooy, *Twins in Contemporary Literature and Culture: Look Twice* (New York: Palgrave Macmillan, 2005), 51–52, 66.

35 twin identity confusion also unsettles us for deeper reasons: I draw here on Nina Strohminger, "The Self Is Moral," *Aeon*, November 17, 2014.

36 "You're a difficult problem": Kafka, *The Castle* (Everyman's Library), trans. Willa and Edwin Muir (New York: Knopf, 1992), 20.

37 "My idea," Twain commented, "is to afford": Mark Twain, *The Prince and the Pauper*, ed. Michael B. Frank and Edward Fischer (Oakland: University of California Press, 2011), xvi.

38 Nigerian hymn: Penelope Farmer, *Two: Or the Book of Twins and Doubles* (London: Virago, 1996), 2.

39 *Breaking Bad:* Strohminger, "Self Is Moral."

40 "I could not bring myself to hate him altogether": Edgar Allan Poe, *Complete Stories and Poems of Edgar Allan Poe* (New York: Knopf, 1966), 160.

40 "You have conquered, and I yield": Ibid., 170.

42 Jack Denfeld Wood and Gianpiero Petriglieri: Jack Denfeld Wood and Gianpiero Petriglieri, "Transcending Polarization: Beyond Binary Thinking," *Transactional Analysis Journal* 35, no. 1 (2005): 31–39.

44 structuralism: A pair of structuralist classics are Ferdinand de Saussure, *Course in General Linguistics*, trans. Wade Baskin (1916; repr., New York: Columbia University Press, 2011), and Claude Levi-Strauss, *Structural Anthropology* (New York: Basic Books, 1974).

44 Hegel argued, more ambitiously, that all: G. W. F. Hegel, *Introduction to the Philosophy of History* (Indianapolis: Hackett, 1988). While he's known for them, Hegel doesn't use the specific terms *thesis*, *antithesis*, and *synthesis* much himself.

44 Janusian thinking: Kat Duff, "Gemini and the Path of Paradox," *Parabola: The Magazine of Myth and Tradition* 19, no. 2 (1994): 16.

44 "two opposing ideas in the mind": F. Scott Fitzgerald, "The Crack-Up," *Esquire*, February 1936, 41.
46 Matthew and Michael Dickman: Rebecca Mead, "Couplet," *New Yorker*, April 6, 2009.
46 Minister for the Exterior: Farmer, *Two*, 45.
48 positive feedback loop: Ian Hacking refers to the general phenomenon as "the looping effect of human kinds" in his book *Rewriting the Soul: Multiple Personality and the Sciences of Memory* (Princeton, NJ: Princeton University Press, 1995), 21. Thanks to Victor Kumar for pointing me to this.
50 "blooming, buzzing confusion": William James, *The Principles of Psychology*, vol. 1 (New York: Henry Holt, 1905), 488.
51 "in that it becomes the mooring of a 'self-system' ": Ricardo Ainslie, *The Psychology of Twinship* (Lincoln: University of Nebraska Press, 1985), 96.
56 The dominant intellectual position: For a helpful, concise historical survey on philosophizing about the self, see John Barresi and Raymond Martin, "History as Prologue: Western Theories of the Self," in *The Oxford Handbook of the Self*, ed. Shaun Gallagher (Oxford: Oxford University Press, 2011), 33–56.
56 it's only a performance, an act: See David Hume, *Treatise of Human Nature* (1739; repr., Oxford: Clarendon Press, 1888), bk. 1, pt. 4, sec. 6.
58 "who we are is to a great extent determined by our friends": Alexander Nehamas, *On Friendship* (New York: Basic Books, 2013), 3.
58 similar to a jazz improvisation: Benjamin Bagley, "Loving Someone in Particular," *Ethics* 125 (January 2015): 480.
59 "always unfinished business": Nehamas, *On Friendship*, 138.
59 Schechtman: Marya Schechtman, *The Constitution of Selves* (Ithaca, NY: Cornell University Press, 2007).

61 One nuanced alternative to seeing twins as binary opposites to see them as polarities: I take this idea from Farmer, *Two*, 332.

62 "the New World prefers intermediate forms": Claude Lévi-Strauss, *The Story of Lynx*, trans. Catherine Tihanyi (Chicago: University of Chicago Press, 1996), 227.

62 "He knows neither good nor evil": Paul Radin, *The Trickster: A Study in American Indian Mythology* (1956; repr., New York: Schecken, 1971), xxiii.

66 "Everything I am not": Michael Dickman, *August 20, 1975* (New York: Kunst Editions, 2009).

67 "This thing of darkness": William Shakespeare, *The Tempest*, ed. Stephen Orgel (Oxford: Oxford University Press, 2008), 202.

68 "The ego keeps its integrity": Carl Jung, "On the Nature of the Psyche" (1954), in *C. G. Jung: The Collected Works*, vol. 8, 2nd ed., ed. H. Read, M. Fordham, and G. Adler (London: Routledge, 1986), 219. See also Carl Jung, "The Archetypes and the Collective Unconscious," in *Collected Works*, vol. 9, pt. 1, 2nd ed. (New York: Bollingen Foundation, 1968), 20–21:

> The meeting with oneself is, at first, the meeting with one's own shadow. The shadow is a tight passage, a narrow door, whose painful constriction no one is spared who goes down to the deep well. But one must learn to know oneself in order to know who one is. For what comes after the door is, surprisingly enough, a boundless expanse full of unprecedented uncertainty, with apparently no inside and no outside, no above and no below, no here and no there, no mine and no thine, no good and no bad. It is the world of water . . .

where I am indivisibly this *and* that; where I experience the other in myself and the other-than-myself experiences me.

2. HOW MANY OF YOU ARE THERE?

73 "How have you made division of yourself?": William Shakespeare, *Twelfth Night*, in *The Pictorial Edition of the Works of Shakespeare*, ed. Charles Knight (George Routledge and Sons, 1867), 180.

73 Nuer: E. E. Evans-Pritchard, "A Problem of Nuer Religious Thought," in *Myth and Cosmos: Readings in Mythology and Symbolism*, ed. J. Middleton (Garden City, NY: Natural History Press, 1967), 134.

73 chief surgeon of the Hôpital Bichat: Hillel Schwartz, *The Culture of the Copy* (New York: Zone Books, 1996), 34.

74 One developmental study: Discussed in Ainslie, *Psychology of Twinship*, 52.

75 A standard assumption of modern Western culture: See Clifford Geertz, "From the Native's Point of View: On the Nature of Anthropological Understanding," *Bulletin of the American Academy of Arts and Sciences* 28, no. 1 (1974).

76 Dreger, who's extensively studied conjoined twins: Alice Domurat Dreger, "The Limits of Individuality: Ritual and Sacrifice in the Lives and Medical Treatment of Conjoined Twins," *Studies in History and Philosophy of Biological and Biomedical Sciences* 29, no. 1 (1998): 9.

81 Katia and Tatiana Hogan: See the CBC documentary *Inseparable: Ten Years Joined at the Head*.

85 "Oh, leave it. Let the maid sweep it up in the morning": Reported by Kierkegaard's former amanuensis, Israel Levin, in Howard V. Hong, "Historical Introduction," in Søren

Kierkegaard, *The Moment and Late Writings* (Princeton, NJ: Princeton University Press, 1998), xxxi.

86 Plato compared reason to a charioteer: Plato, *Phaedrus* 246b, in *Plato: Complete Works*, 524.

88 Wari people: See Beth Conklin and Lynn Morgan, "Babies, Bodies and the Production of Personhood in North American and Native Amazonian Society," *Ethos* 24, no. 4 (1996): 657–94.

89 "Twins cannot read each other's minds": Michael Shermer, foreword to Segal, *Twin Mythconceptions*, xi. Segal discusses the evidence on pages 143–50.

91 Andy Clark and David Chalmers: Andy Clark and David Chalmers, "The Extended Mind," *Analysis* 58, no. (1998), 7–19. Some theorists have argued that not just your mind (as Clark and Chalmers argue) but also your *body* can extend beyond your skin. (See Margrit Shildrick, "'Why Should Our Bodies End at the Skin?': Embodiment, Boundaries and Somatechnics," *Hypatia* 30, no. 1 (2015): 13–29.) The most obvious examples are people with disabilities who use material technologies, such as limb prostheses, supports for sensory functions, wheelchairs, ventilators, behavior-controlling drugs, neuro-implants, cochlear implants, and transplanted organs. But some argue that, in our technologically advanced world, "the prostheticized body is the rule, not the exception" (David Mitchell and Sharon Snyder, *Narrative Prosthesis: Disability and the Dependencies of Discourse* [Ann Arbor: University of Michigan Press, 2000], 7).

92 transactive memory: The idea first appeared in D. M. Wegner, T. Giuliano, and P. Hertel, "Cognitive Interdependence in Close Relationships," in *Compatible and Incompatible*

92   *Relationships*, ed. W. J. Ickes (New York: Springer-Verlag, 1985), 253–76.

92   "Googling *each other*": Joshua Wolf Shenk, *Powers of Two: Finding the Essence of Innovation in Creative Pairs* (New York: Eamon Dolan / Houghton Mifflin Harcourt, 2014), 49.

95   Tim and Greg Hildebrandt: Harvey Stein and Ted Wolner, *Parallels: A Look at Twins* (New York: Dutton, 1978).

95   offered by Bennett Helm: Bennett W. Helm, *Love, Friendship and the Self: Intimacy, Identification and the Social Nature of Persons* (Oxford: Oxford University Press, 2012).

95   Aristotle had in mind: Aristotle, *Nicomachean Ethics*, in *Complete Works of Aristotle: The Revised Oxford Translation*, vol. 2, ed. Jonathan Barnes (Princeton, NJ: Princeton University Press, 2014), 1166a30–33, 1166b1, 1169b6, 1170b6f, 1171b33 / 1843, 1848, 1850.

96   "in the best instances . . . absolute mutual trust": Spinney, "Twin Boom."

96   "I was already so formed": Montaigne, "Of Friendship," in *The Complete Essays of Montaigne*, trans. Donald Murdoch Frame (Stanford, CA: Stanford University Press, 1958), 143.

102   an important theme in feminist theory: The literature is vast, but one good place to start is Eva Feder Kittay, "The Ethics of Care, Dependence and Disability," *Ratio Juris* 24, no. 1 (2011): 49–58.

102   "The individual does not oppose himself": Georges Gusdorf, "Conditions and Limits of Autobiography," trans. James Olney, in James Olney, *Autobiography* (Princeton, NJ: Princeton University Press, 1980). See also James G. Carrier, "People Who Can Be Friends: Selves and Social Relationships," in *The Anthropology of Friendship*, ed. Sandra Bell and Simon Coleman (London: Routledge, 2020).

104 a truth not just about ourselves, but about everyone: See Kittay, "Ethics of Care," 51: "The emphasis on independence extols an idealization that is a mere fiction, not only for people with disability, but for all of us."

104 "It's high time you quit being twins": Carol Morse, *Double Trouble* (Garden City, NY: Doubleday, 1964). See also the back cover of Catherine Storr's *Puss and Cat* (New York: Penguin, 1978): "At first being identical twins was fun, but in the end Puss and Cat decided they wanted to be people in their own right."

108 "the smallest indivisible human unit is two people": Tony Kushner, in Shenk, *Powers of Two*.

108 "Neither of us reserved anything for himself": Montaigne, "Of Friendship," 139.

108 Romantic relationships between men and women: See Marilyn Friedman, "Romantic Love and Personal Autonomy," *Midwest Studies in Philosophy* 22, no. 1 (1998): 162–81.

108 "husband and wife are one, and that one the husband": Susan B. Anthony, "Homes of Single Women," in *The Elizabeth Cady Stanton–Susan B. Anthony Reader: Correspondence, Writings, Speeches*, ed. Ellen Carol DuBois (1877; repr., Boston: Northeastern University Press, 1992), 134. The English jurist William Blackstone explained the legal tradition of "coverture" Anthony was referring to as follows: "By marriage, the husband and wife are one person in law: that is, the very being or legal existence of the woman is suspended during the marriage, or at least is incorporated and consolidated into that of the husband: under whose wing, protection, and *cover*, she performs every thing . . . a man cannot grant any thing to his wife, or enter into covenant with her: for the grant would suppose her separate existence; and to covenant with her, would be only to

covenant with himself." Sir William Blackstone, *Commentaries on the Laws of England* (Philadelphia: George T. Bisel, 1922), bk. 1, chap. 15, 441–42.

109 "half people": On women as "half people," see Rebecca Traister, *All the Single Ladies: Unmarried Women and the Rise of an Independent Nation* (New York: Simon and Schuster, 2016), 26.

109 views the liberal attachment: On the importance of the separateness of persons to feminism, see Martha Nussbaum, *Sex and Social Justice* (Oxford: Oxford University Press, 1999), 59–67, "The Feminist Critique of Liberalism."

110 Alice Dreger notes that conjoined twins: Alice Domurat Dreger, *One of Us: Conjoined Twins and the Future of Normal* (Cambridge, MA: Harvard University Press, 2004).

110 Chang and Eng Bunker: Grosz, "Intolerable Ambiguity," 62.

3. ARE YOU TWO IN LOVE?

116 "Even if you're unsure of God": Stein and Wolner, *Parallels*.

117 in the creation myth: Origin myths of humans as divided or untwinned creatures appear in other cultures, too. According to Hawaiian tradition, when the first human was ordered to earth, only half of him was delivered, and the remainder of the package will only appear at the end of the world (Farmer, *Two*, 333). The Yoruba tribe of West Africa believe, similarly, that we all have a double residing in heaven—unless we're twins, in which case we get to enjoy our counterpart down here on earth (ibid., 365). The Jewish mystical text the *Zohar* claims, "Each soul and spirit prior to its entering into this world, consists of a male and female united into one being. When it descends on this earth the two parts separate and animate two different bodies" (Christian D.

Ginsburg, *The Kabbalah: Its Doctrines, Development and History* [London: Longmans, Green, Reader and Dyer, 1865], 34). In Islam, the prophet Muhammad is said to have referred to women as "the twin halves of men" (H. Jawad, *The Rights of Women in Islam* [London: Palgrave Macmillan, 1998], 7).

118 "And when a person meets the half that is his very own": Plato, *Symposium*, in *Plato: Complete Works*, 475.

118 "one soul of interwoven flame": Percy Bysshe Shelley, *The Poetical Works* (London: E. Moxon, 1884), 213, "Rosalind and Helen."

118 "Surely you and everybody have a notion": Emily Brontë, *Wuthering Heights* (London: Heron Books, 1966), 84.

118 the core features of our contemporary ideal of romantic love: Simon May, *Love: A History* (New Haven, CT: Yale University Press, 2012).

119 "What were the use of my creation": Brontë, *Wuthering Heights*, 84.

119 as Alice Dreger points out: Dreger, *One of Us*, 50. "I've Got You Under My Skin," music and lyrics Cole Porter, 1936; copyright The Cole Porter Trusts, New York, NY. I recently came across a greeting-card inscription—quoting Jackson Nieuwland's *I am a human being* (Compound Press, 2020)—that made the analogy explicit: "We're conjoined twins. The right lobe of your brain is the left lobe of mine."

119 "cannot say what it is they want from one another": Plato, *Symposium*, in *Plato: Complete Works*, 475.

120 Castor and Pollux: Technically, the pair aren't twins, since they have different dads. To put it more precisely, they are twins, but not *each other's*: Pollux was said to come from the same egg as Helen of Troy (fertilized by Zeus), Castor from the same egg as Clytemnestra (fertilized by Tyndareus),

both of which were laid by Leda simultaneously. (A celebrity family to rival the Kardashians.) Whatever the biology, the brothers functioned culturally as twins from the outset. They were referred to as Gemini (the Latin for "twin") in ancient Rome, and they've been known as the Divine Twins in discussions of mythology for centuries.

121 According to the ancient Greek poet Pindar: Pindar, *Nemean Ode* 10, in *Nemean Odes. Isthmian Odes. Fragments*, ed. and trans. William H. Race (Cambridge, MA: Loeb Classical Library, Harvard University Press, 1997), line 89.

121 "If all else perished": Brontë, *Wuthering Heights*, 84.

128 I fell in love with a woman myself: In case you're wondering, which you are, Nancy Segal reports, "If one identical twin is homosexual, then his or her cotwin has a higher probability of being, or becoming, homosexual relative to a fraternal twin or nontwin relative. However, the chance is far less than 100%" (Segal, *Twin Mythconceptions*, 188).

129 "of a surpassingly beautiful curve": Edgar Allan Poe, "The Fall of the House of Usher," in *Complete Stories and Poems*, 180.

130 "Their eyes met," "no more masculine," "self-absorbed invalids," "childlike despite their nineteen years": Thomas Mann, "Blood of the Walsungs," in *Death in Venice and Other Tales*, trans. Joachim Neugroschel (New York: Penguin, 1999), 261, 257, 268, 256. Though I'm focusing on the homophobic stereotypes here, the story trades heavily in antisemitic stereotypes, too. In an extra layer of creepiness, Mann's wife, Katia, was both Jewish and a different-sex twin.

131 "transitional objects": The concept first appeared in D. W. Winnicott, "Transitional Objects and Transitional Phenomena: A Study of the First Not-Me Possession," *International Journal of Psychoanalysis* 34 (1953): 89–97.

- 131 arrested development: See Martin Hoffman, *The Gay World: Male Homosexuality and the Social Creation of Evil* (New York: Basic Books, 1968).
- 131 Freud viewed same-sex desire: Sigmund Freud, "'Civilized' Sexual Morality and Modern Mental Illness," *The Standard Edition of the Complete Psychological Works of Sigmund Freud*, vol. 9 (London: Hogarth Press, 1908), 200.
- 131 "An effeminate homosexual": Quentin Crisp, *The Naked Civil Servant* (New York: Holt, Rinehart and Winston, 1968), 130.
- 132 "pla[y] like little puppies," "bundled up so warmly," "spoke as mindlessly": Mann, "Blood of the Walsungs," 270, 279, 261.
- 132 twincest: The fixation appears across the planet. Javanese tradition assumes that all young different-sex twins are on track for sleeping with each other; in nearby Bali they're thought to have already managed it in utero and were once killed as a precaution. Some Romani groups actively encourage different-sex twins to marry or to make out in public (Piontelli, *Twins in the World*, 186–88).
- 132 twincest is rare: See Geoff Puterbaugh, *Twins and Homosexuality: A Casebook* (New York: Garland, 1990), xiv; Lawrence Wright, *Twins: Genes, Environment and the Mystery of Identity* (London: Weidenfeld and Nicolson, 1997), 46.
- 133 stories about queer couples have almost never ended happily: One tweeter writes, "Kudos for including such a well-developed gay character! Have you figured out how you're going to kill them yet?" @WorstMuse, Twitter, June 20, 2016.
- 133 "she is perhaps more like Alice than she ever realized": "Think Twice," Lisa Scottoline, accessed January 12, 2023, https://www.scottoline.com/books/rosato-associates-series/think-twice/.

134 "Twins who aren't separated early . . . are frequently doomed": de Nooy, *Twins in Contemporary*, 22. Chapter 2 of this book, "Twins and the Couple: Surviving Sameness in Novels of Twin Lives," informed much of my early thinking about the association between twins and queerness.

136 "Whatever our differences . . . we have historically been regarded . . . as twins": Joseph Bristow, *Sexual Sameness: Textual Differences in Lesbian and Gay Writing* (Abingdon, U.K.: Routledge, 1992), 3.

138 "romantic friendship": See Sukaina Hirji and Meena Krishnamurhty, "What Is Romantic Friendship?" *New Statesman*, November 2, 2021.

140 we're all worried we're a little bit twinny: In one version of the Greek myth, Narcissus gazed into the pond not because he was in love with himself, but because his reflection reminded him of his beloved lost twin sister (Schwartz, *Culture of the Copy*, 17). To a certain kind of singleton, that might not be much of an improvement.

140 "Methinks you are my glass": William Shakespeare, *A Comedy of Errors* (Oxford: Oxford University Press, 2003), 180.

141 "not because he's handsome": Brontë, *Wuthering Heights*, 82.

141 love is "a sort of secession": C. S. Lewis, *The Four Loves* (San Francisco: HarperOne, 2017).

141 marked impression of self-sufficiency: A more specific fear of secession is stoked by female twins in particular. As Alison Pingree writes in an article on the conjoined vaudeville twins Daisy and Violet Hilton, the sight "of two young women who go everywhere together—with no man in sight—and thoroughly enjoy themselves" represents a major threat to the patriarchy. Pingree writes of the Hiltons, who

performed in the 1920s and '30s, "The twins' particular form of aberration perfectly embodies what many by then had come to fear: that a woman's body might not be able to be controlled; that heterosexual, companionate marriage might not be the only form of intimate 'bonding' between two people." Pingree, " 'Exceptions That Prove the Rule.' "

141 "about the self that is at odds with everything around it": bell hooks, in a panel conversation in 2014 at the New School called "Are You Still a Slave? Liberating the Black Female Body."

142 amatonormativity: Elizabeth Brake, *Minimizing Marriage: Marriage, Morality and the Law* (New York: Oxford University Press, 2012).

142 "Who d'you love Most in the World?": Arundhati Roy, *The God of Small Things* (Toronto: Random House of Canada, 1997), 144.

143 identical twins are less likely to marry: Piontelli, *Twins in the World*, 18. Studies suggest that twins who do get married are no more likely than non-twins to divorce, though. (The divorce rate for *parents* of twins is above average.) See Segal, *Twin Mythconceptions*, 214–16.

143 Deena Lilygren: "I Love Living with My Best Friend, So I Bought a House with Her and Her Husband," *Huffington Post*, August 14, 2019.

144 "we have very little access to each other's interior world": Rebecca Nicholson, "Tegan and Sara: People Never Talk About Women and Drug Use Positively" (interview), *Guardian*, September 18, 2019.

144–45 "Our shared best friends acted as a conduit": Sara Quin and Tegan Quin, *High School* (New York: Simon and Schuster, 2020), 24.

145 "Kristina punched Karissa": "*Playboy*'s Shannon Twins Knock Each Other Silly in Brawl," *TMZ*, May 27, 2017.

145 Jeena Han: Kelly Puente, "'Evil Twin' Granted Parole After Serving Nearly Two Decades in Prison for Plot to Kill Sister in Irvine," *Orange County Register*, November 27, 2017.

145 Real-life twins don't generally "turn out to be mentally defective": Piontelli, *Twins in the World*, 18.

145 lower risk of depression and suicide: Twins are about 25 percent less likely than non-twins to take their own lives, according to a Danish study. The statistic held for both male and female and MZ and DZ twins. See Cecilia Tomassini, Knud Juel, Niels Vilstrup Holm, Axel Skytthe, and Kaare Christensen, "Risk of Suicide in Twins: 51-Year Follow-up Study," *BMJ Clinical Research* 327, no. 7411 (2003): 373–74.

146 significant lead over singletons is grief: Studies have shown that when one twin dies, the surviving twin's score on the Grief Experience Inventory is, on average, the highest on the planet. See N. L. Segal, S. M. Wilson, T. J. Bouchard, and D. G. Gitoin, "Comparative Grief Experiences of Bereaved Twins and Other Bereaved Relatives," *Personality and Individual Differences* 18, no. 4 (1995): 511–24.

149 "That's more, in one minute, lavished on that little, senseless oyster": Anne Brontë (writing as Acton Bell), *The Tenant of Wildfell Hall* (London: T. C. Newby, 1848), 2:155.

4. HOW FREE ARE YOU?

157 Francis Galton noted the scientific usefulness: Francis Galton, "The History of Twins, as a Criterion of the Relative Powers of Nature and Nurture," *Fraser's Magazine* 12 (1875): 566–76.

158 "the Rosetta Stone of behavioral genetics": Thomas Bouchard, foreword to Nancy Segal, *Entwined Lives: Twins*

and What They Tell Us About Human Behavior (New York: Dutton, 1999), ix.

158 "heritability" estimates: For a summary, see Matt Ridley, *Nature via Nurture: Genes, Experience, and What Makes Us Human* (New York: HarperCollins, 2003), 82–92. On personality similarities, see Segal, *Twin Mythconceptions*, 253.

160 Galton's doubts that "nurture can do anything at all": Sir Francis Galton, *Inquiries into Human Faculty and Its Development* (New York: Macmillan, 1883), 240.

161 the debate over metaphysical freedom: For a good concise introduction, see Thomas Pink, *Free Will: A Very Short Introduction* (Oxford: Oxford University Press, 2004).

163 Compatibilists argue: For a detailed discussion of the many variants, see sections 2.2–2.4 of Timothy O'Connor and Christopher Franklin, "Free Will," in *The Stanford Encyclopedia of Philosophy*, Winter 2022 ed., ed. Edward N. Zalta and Uri Nodelman.

164 Compatibilists have made a lot of ingenious moves: One popular strategy is to say that you're free when you act not just on any old desire, but on the desires you desire to have (or, as it's sometimes put, the desires you identify with or endorse). On this view, I may want to text my ex, but I don't *want* to want to text the ex, and that explains why my act is unfree. In cases where I do endorse my desires, I act freely, even if my endorsements are themselves determined. For the original statement of this idea, see Harry Frankfurt, "Freedom of the Will and the Concept of a Person," *Journal of Philosophy* 68, no. 1 (1971): 5–20.

167 "petty word-jugglery" and "a wretched subterfuge": Kant, *Critique of Practical Reason*, trans. Thomas Kingsmill Abbott (London: Longmans, Green, 1889), 189.

168–69 "extremely damaging to our view of ourselves": Derk Pereboom, "Free-Will Skepticism and Meaning in Life," in *The Oxford Handbook of Free Will*, 2nd ed., ed. Robert Kane (New York: Oxford University Press, 2011), 420.

169 "with more than Baron Munchhausen's audacity": Friedrich Nietzsche, *Beyond Good and Evil: Prelude to a Philosophy of the Future*, trans. Walter Kaufman (1889; repr., New York: Knopf, 1989), 28.

169 "literal self-creation is not just empirically, but logically impossible": Susan Wolf, "Sanity and the Metaphysics of Responsibility," in *The Philosophy of Free Will: Essential Readings from the Contemporary Debates*, ed. Paul Russell and Oisin Deery (Oxford: Oxford University Press, 2013), 288.

170 Sometimes I trace this take to my being a woman, or someone with a physical disability: Or maybe to my nationality? When I raised this take on free will with my Canadian philosophy colleague, she wrote, "I always thought of it as a kind of American thing. Lol."

171 Such twins are less likely to be invested in individualism: It's important to acknowledge the diversity of twin experiences here. One twin who has often found her twin dominating told me, "I can report being *extremely* invested in individualism."

173 Oskar Stohr and Jack Yufe: See Nancy L. Segal, *Indivisible by Two: Lives of Extraordinary Twins* (Cambridge, MA: Harvard University Press, 2005), 50–78.

173 "genealogical anxiety": Amia Srinivasan, "Genealogy, Epistemology and Worldmaking," *Proceedings of the Aristotelian Society* 119, no. 2 (2019): 127–56, at 128.

177–78 "Part of my reluctance about having a twin": Elyse Schein and Paula Bernstein, *Identical Strangers: A Memoir*

of *Twins Separated and Reunited* (New York: Random House, 2008), 47.
178 "The vision that scares me the most": Ibid., 12.
181 "all those unfulfilled but possible futures": Sigmund Freud, *Collected Papers*, vol. 4 (London: Hogarth Press, 1925), 388.
181 "I would have, if I'd been my twin": Albert Goldbarth, "Alien Tongue," *Northwest Review* 27, no. 3 (1989): 34.
181 it makes perfect sense to feel a sense of loss: Kieran Setiya, *Midlife: A Philosophical Guide* (Princeton, NJ: Princeton University Press, 2017), chap. 3.
182 "the fact that there are so many different things worth wanting": Ibid., 60.
182 the devil would be in the details: Ibid., chap. 4.
183 "It is rational to respond more strongly to the facts": Ibid., 101.
183 "An actual life is not a thought experiment": Ibid., 87.
183 "If you rewound the tape and started anew": "Can Progressives Be Convinced That Genetics Matters?" *New Yorker*, September 13, 2021.

5. WHAT ARE YOU FOR?

195 "the twin mystique": Joan A. Friedman, *The Same but Different: How Twins Can Live, Love and Learn to Be Individuals* (Los Angeles: Rocky Pines Press, 2014), 11.
196 childhood fantasy of having a twin: Dorothy T. Burlingham, "The Fantasy of Having a Twin," *Psychoanalytic Study of the Child* 1, no. 1 (1945): 205–10.
197 "I am small and weak in the face of dangers": Ibid., 210.
197 "One of these men is *genius* to the other": Shakespeare, *Comedy of Errors*, 175.
197 singletons each have a "shadow self": Jung, "Archetypes and the Collective Unconscious," 20–22.

199 Nussbaum distinguishes a menu: Martha Nussbaum, "Objectification," *Philosophy and Public Affairs* 24, no. 4 (1995): 249–91; the list of features is at 257.

199 Langton adds three more moves: Rae Langton, *Sexual Solipsism: Philosophical Essays on Pornography and Objectification* (Oxford: Oxford University Press, 2009), 228–29.

199 "There are two of them and they are the same": Shari Waxman, "The Twins Thing," *Salon*, May 30, 2003.

200 Mengele's experiments: Lucette Matalon Lagnado and Sheila Cohn Dekel, *Children of the Flames: Dr. Josef Mengele and the Untold Story of the Twins of Auschwitz* (New York: Penguin, 1992).

201 freak show: For illuminating discussions, see Garland-Thomson, *Freakery* and *Extraordinary Bodies*; Adams, *Sideshow USA*; and Durbach, *Spectacle of Deformity*.

203 genes or highly specific aspects of our bodies: Dona Lee Davis, *Twins Talk: What Twins Tell Us About Person, Self and Society* (Athens: Ohio University Press, 2015), 12, 92.

203 Films rarely provide: For a host of examples of objectifying uses of twins in film, entertainment, and advertisements, see de Nooy, *Twins in Contemporary*, Hillel Schwartz, *Culture of the Copy*, and Karen Dillon, *The Spectacle of Twins in American Literature and Popular Culture* (Jefferson, NC: McFarland, 2018).

205 positively valuable: Nussbaum, "Objectification," 265, 290.

206 not being physically objectified enough: Leslie Green, "Pornographies," *Journal of Political Philosophy* 8, no. 1 (2000): 27–52, at 45–46.

207 deciding factor in assessing objectification: Nussbaum, "Objectification," 275.

207 "nobody [sees] any differences between twins and goats": Piontelli, *Twins in the World*, 153.

208 female nudes: Anne Eaton, "What's Wrong with the (Female) Nude?: A Feminist Perspective on Art and Pornography," in *Art and Pornography: Philosophical Essays*, ed. Hans Maes and Jerrold Levinson (Oxford: Oxford University Press, 2012), 277–308.

209 "Declaration of Rights and Statement of Need of Twins and Higher Order Multiples": Council of Multiple Birth Organisations.

209 Ann Cahill: Ann J. Cahill, *Overcoming Objectification* (Abingdon, U.K.: Routledge, 2010), chap. 2.

213 Lexie and Allie Kaplan: Dillon, *Spectacle of Twins*, 56–58.

214 "I guess just the goal in general": Paige Skinner, "Brooklyn and Bailey McKnight Are YouTube's Big Sisters," *Dallas Observer*, April 9, 2019.

214 Chang and Eng Bunker: For an engaging retelling of the Bunkers' story, see Yunte Huang, *Inseparable: The Original Siamese Twins and Their Rendezvous with American History* (New York: Liveright, 2018).

214 Bogdan has argued: Robert Bogdan, *Freak Show: Presenting Human Oddities for Amusement and Profit* (Chicago: University of Chicago Press, 1990).

214 "the spectacle of twinship": Dillon, *Spectacle of Twins*, 2.

216 Andrea Dworkin: Andrea Dworkin, *Our Blood: Prophecies and Discourses on Sexual Politics* (New York: Harper and Row, 1976), 78.

216 Sandra Bartky: Sandra Bartky, "Foucault, Femininity, and the Modernization of Patriarchal Power," in *Femininity and Domination: Studies in the Phenomenology of Oppression* (New York: Routledge, 1990), 78.

216 Catharine MacKinnon: Catharine A. MacKinnon, *Toward a Feminist Theory of the State* (Cambridge, MA: Harvard University Press, 1989), 136–37.

217 "Women live in objectification": Catharine A. MacKinnon, "Sexuality, Pornography and Method: 'Pleasure Under Patriarchy,'" *Ethics* 99 (1989).

217 in the late 2000s: For a concise and helpful recent introduction to the history of feminism, see Carol Hay, *Think Like a Feminist: The Philosophy Behind the Revolution* (New York: Norton, 2020).

218 "Words or images that literally depict a woman's powerlessness": Nadine Strossen, *Defending Pornography: Free Speech, Sex, and the Fight for Women's Rights* (New York: Scribner, 1995), 174.

218 christened "freak studies": Michael M. Chemers, "Staging Stigma: A Freak Studies Manifesto," *Disability Studies Quarterly* 25, no. 3 (2005).

223 opening desire up for political critique: Amia Srinivasan, *The Right to Sex* (New York: Farrar, Straus and Giroux, 2021).

224 Zazzo reports a similarly depressing: René Zazzo, *Le paradoxe des jumeaux*, cited in Farmer, *Two*, 161.

CODA

230 "The cultural work done in the past by gods": Barthes, *Mythologies* (New York: Farrar, Straus and Giroux, 1972).

231 "discrete bubbles floating past each other": Helen Garner, *Joe Cinque's Consolation: A True Story of Death, Grief and the Law* (Sydney: Pan Macmillan, 2004), 177.

A NOTE ON THE AUTHOR AND ILLUSTRATOR

HELENA DE BRES is a professor of philosophy at Wellesley College, where she researches and teaches ethics, philosophy of literature, and political theory. Her essays and humor writing have appeared in *The Point*, the *New York Times*, the *Los Angeles Review of Books*, *McSweeney's Internet Tendency*, and elsewhere. Her book *Artful Truths: The Philosophy of Memoir* was published by the University of Chicago Press in 2021. She lives in Somerville, Massachusetts.

JULIA DE BRES is a freelance illustrator and a senior lecturer in linguistics at Massey University. She analyzes how minority groups use language to resist social inequalities and illustrates the results of her research. She lives in Wellington, New Zealand.